Perry Goldman

Build a Successful Sales Program

A Service of

NAHB

BuilderBooks™
National Association of Home Builders
1201 15th Street, NW
Washington, DC 20005-2800
www.builderbooks.com

Build a Successful Sales Program
Perry Goldman

Doris Tennyson	Senior Acquisitions Editor
Jenny Stewart	Assistant Editor
Karen Ruppert	Copyeditor
Armen Kojoyian	Cover Designer

BuilderBooks at the National Association of Home Builders

ERIC JOHNSON	Publisher
THERESA MINCH	Executive Editor
DORIS TENNYSON	Senior Acquisitions Editor
JESSICA POPPE	Assistant Editor
JENNY STEWART	Assistant Editor
BRENDA ANDERSON	Director of Fulfillment
GILL WALKER	Marketing Manager
ACQUELINE BARNES	Marketing Manager

GERALD HOWARD	NAHB Executive Vice President and CEO
MARK PURSELL	Executive Vice President Marketing & Sales
GREG FRENCH	Staff Vice President, Publications and Affinity Programs

ISBN 0-86718-561-9

Printed in the United States of America

Cataloging-in-Publication Data available at the Library of Congress

Disclaimer
This publication is designed to provide accurate and authoritative information in regard to the subject matter covered. It is sold with the understanding that the publisher is not engaged in rendering legal, accounting, or other professional service. If legal advice or other expert assistance is required, the services of a competent professional person should be sought.
—From a Declaration of Principles jointly adopted by a Committee of the American Bar Association and a Committee of Publishers and Associations.

For further information, please contact:
BuilderBooks™
National Association of Home Builders
1201 15th Street, NW
Washington, DC 20005-2800
(800) 223-2665
Check us out online at: www.builderbooks.com

10/03 [Armen Kojoyian]/[SLR Production]/[DRC] 2000

ABOUT THE AUTHOR

Perry Goldman, MIRM, is president of Westchester Condos & Coops and New Home Sales & Marketing, Croton-on-Hudson, New York. WCC is the oldest continuously operating condominium and cooperative firm in New York state. New Home Sales & Marketing provides consulting services, strategic marketing plans, and trained onsite sales staffs for new home builders. He received his Ph.D. from Columbia University and has almost a quarter century experience in working with builders, developers, and sponsors of new communities throughout the country. He has also served in a variety of positions with leading advertising and public relations agencies and real estate brokerage firms.

Dr. Goldman has trained hundreds of resale agents and sales counselors for onsite work. He is a New York state real estate instructor, a licensed New York and New Jersey real estate broker, and holds many coveted real estate designations. He has written several articles and conducted seminars on a variety of topics. He is a member of the institute of Residential Marketing, the National Association of Home Builders, and the Builders Association of the Hudson Valley.

ACKNOWLEDGMENTS

I have been privileged to work with many outstanding builder-developers and sales counselors and would like to acknowledge them (in alphabetical order) for their contributions to my growth and development. The former include William Balter, Martin Berger, James Brooks, Robert Hankin, and Richard and Ronald Steinberg. Sales colleagues include Judy Brackenrig, Donna Dencker, Julie Ann Farrell, Linda Lange, Virginia Maus, Karinn Nazak, and Barbara Prackelt.

Without the encouragement and support of my wife, Renee, a distinguished realtor, and our sons Brian and Jared, this volume would not have been written.

The author wishes to acknowledge the efforts of a dedicated and perspicacious editorial staff for their advice, encouragement, and expertise that has guided this book to the marketplace.

BuilderBooks expresses its appreciation to Margaret Meyer for reviewing *Build a Successful Sales Program*.

CONTENTS

FIGURES

INTRODUCTION

If you fail to plan, you're planning to fail

This book offers a compact, comprehensive, and systematic compilation of field-tested materials essential to organize and administer an efficient and effective community sales program. It provides a variety of forms that sales and administrative personnel will find invaluable in managing the countless details in the everyday routine of the home selling process. Each of the forms, from the prospect information card to the new homeowner's satisfaction survey, will yield significant insights if used and analyzed regularly. These forms will also streamline tasks, speed processes, pinpoint areas that require attention, and promote a professional image.

Novices can use these forms to develop a community sales program. Experienced professionals can select those that are applicable but different from ones currently in use and add, delete, or modify the content to reflect their own companies' methods, policies, and procedures. The forms can be copied or customized and used independently of one another. They provide a complete paper trail that guides, monitors, and traces the steps in each individual sales transaction. Collectively, the forms constitute a meticulously documented history of the community's entire sales program.

Chapters One and Two provide an overview of the book's contents followed by a description and an explanation of the system and a discussion of its purpose and how it fits into the other systems. Next, each form is introduced with its description, purpose, and use, its preparation, organization, revision, transmission, and filing. Five of the remaining six chapters present forms that deal with a single step in the sales process—Selections, Financing, Closing Procedures, Post-Closing Purchaser Contact, and Competitive Communities. Chapter Five presents collateral and special use items. The format in the last six chapters will vary somewhat from that of the first two due to the nature of the material. Throughout the book the forms are introduced in the order in which they are most likely to be used with prospects and purchasers. Examples of most of the forms can be found in the chapter in which they were introduced.

Once the user of this book has selected and suitably customized the forms for his/her particular community, he/she will have devised a standardized system to control the flow of information and paperwork and will have created an invaluable resource and reference for the conduct of a successful community sales program.

HOW TO USE THIS BOOK AND THE DISK

The book is written from the perspective of a medium-sized production builder who (a) uses an in-house sales and administrative staff or employs a sales and marketing company in an onsite sales center; (b) offers a generous selection of pre-priced options and upgrades; and (c) allows some degree of custom changes to the company's floor plans and specifications. Of course, small-volume builders with a less-extensive agenda and more restrictive policies may selectively use the forms.

The word "form" is used in the generic sense to encompass a diversity of materials, including an agreement, card, chart, checklist, letter, list, "protocol" (procedural guideline or policy), report, spreadsheet, survey, or worksheet.

To revise the content using the disk that came with this book:

◆ Begin with the commands to "search and replace" and incorporate your company name where the book says *builder's company name*. You might insert your company's forms in place of those included in the book.

◆ Delete entries that do not apply to your company.

◆ Change forms as necessary to reflect your company's policies or procedures. Adjust terms to reflect the legal requirements and customs in your area.

Throughout the book, boxes like this one will provide helpful hints and suggestions.

◆ Add forms that address the specific policies and procedures relevant to the way you do business.

These boxes will not appear when you print your forms. They appear only in the printed copy of the book to help you think through the policy and procedure.

The starting point is with the first prospect to enter the sales center. Everything and everyone has been properly prepared and the community is ready for the marketplace. Signage is in place, the sales center is appropriately designed and outfitted, the model merchandising is complete, brochures are collated, the public relations and advertising campaigns are launched, and directions to the community have been placed next to each phone. The sales staff has been engaged, company policies and procedures have been reviewed, and orientation with the builder and construction staff has been completed. Finally, with forms that have been suitably crafted, we are now well prepared and properly positioned for the conduct of a successful community sales program.

CHAPTER 1

ONSITE INSIGHT

The onsite specialists may be either independent contractors or employees who are "married" to their builder's product. They have to sell (only) what they have to sell, and although they may see many more prospects than the resale agent does, a far smaller percentage actually purchase their homes—only three percent to five percent based on widespread experience. Initially, they have far less control over prospects who arrive in their own vehicles (often simultaneously with other prospects) on most sites without prior appointments. Depending on the stage of the new development, conditions (during the initial "pre-construction" period) may be less than optimum with only blueprints to work from and a "dream" to sell. Eventually, the "Grand Opening" will feature a well-equipped sales center with attractive graphics displays illustrating the elevations and various aspects of the new community and, if fortunate, one or more merchandised model homes. To master their craft, onsite sales counselors must adroitly use the available sales aids, diligently repeat routines and activities, and carefully oversee a vast array of paperwork. They must be prepared with answers to the most frequently asked questions and be familiar with the differences in forms of homeownership. (These items are addressed in Appendices A through F.) They must remain focused and unflappable through the quiet times of few visitors and phone calls and through the hectic, barely controllable times of multiple and simultaneous demands for their attention. Although the focus of this book is on setting up and administrating a successful community sales program, the requisite attributes for a successful onsite sales counselor deserve additional consideration.

First and foremost of these attributes is the ability to communicate clearly, concisely, confidently, and comfortably and to listen carefully. This in turn is predicated upon acquiring detailed construction and product knowledge (both your own and that of your competitors), observing the proper sequence of the sales process (sometimes referred to as the "Critical Path" or "Power Process"), and understanding the personality types and styles of prospects and the preferred ways of responding to each. The onsite sales counselor also interacts regularly with advertisers, appraisers, architects, attorneys, the builder, construction personnel, investors, lenders, model merchandisers, mortgage officers, public relations people, Realtors, suppliers, tradesmen, and vendors. It's imperative that one be articulate and personable—in short, be able to achieve rapport with diverse groups of people, each with their own agenda but each integral to your success.

Second and no less important but often lacking is detail-oriented administrative and organizational talent. We address this deficiency by providing samples of those items—from the prospect information (registration) card to the new homeowner's satisfaction survey to agreements, charts, checklists, files, letters, lists, protocols (procedural guidelines), reports, spreadsheets, and worksheets—that onsite specialists can customize to the requirements of their community. Confident that they have virtually everything in place that they need to manage the sales processes, they can then concentrate on "sales" and their professional development.

Finally, a word about the need for emotional stamina and self-renewal. It's not easy to adjust to a controlled environment with set hours and be effective day after day in a limited area of operations. Additionally, the "irregular" schedule of working weekends regularly, the vagaries of the marketplace, the almost inevitable and uncontrollable delays seemingly inherent in new construction, and the personality changes and stresses that sometimes surface among buyers makes one readily appreciate the need to be cool, calm, collected, and positive in trying circumstances. Onsite sales can be a draining and at times lonely experience.

Conversely, it's extremely gratifying to see a family glow with pride as they move into their new home, knowing that you helped make it happen!

OVERVIEW

The successful organization and administration of a community sales program requires a set of basic forms to facilitate a seamless transition from the initial prospect encounter to the final settlement or transfer of title at closing. Astute onsite sales counselors will, of course, utilize all of the displays and aids available in the sales environment to make prospects feel comfortable, informed, interested, and excited about their community. They will need a variety of forms to control and manage the sales process, beginning with visitor welcome and guest registrations. The use of standardized forms will ensure consistency and uniformity in policies and procedures, control the flow of paperwork, and document each successive step in the sales process. If the forms are filled out completely and filed properly, they will not only expedite the sales process but also prevent costly delays and embarrassment caused by the loss of vital information.

DESCRIPTION AND EXPLANATION

All too often we have seen sales information jotted down on scraps of paper, usually undated and unsigned, and simply thrown haphazardly and loosely into files or folders. Our forms are designed to obtain vital information and be filed, sorted, and stored in designated locations so that sought-after data is easily retrievable and readily accessible to anyone who may need it. The forms are presented in the order in which they are most likely to be used. This order allows the sales staff to proceed with their prospects in a logical sequence throughout the investment process. Each form will be described with its purpose and use explained, including who prepares it, who updates it, who gets copies, how it is transmitted, and who files it and where. A chart conveniently summarizes this material (see Form List). The forms constitute an oversight system that helps to pinpoint problems, assess progress, and monitor profits. The forms also provide a paper trail that will prove useful in instances of misunderstandings or disagreements with purchasers or even future litigation. The forms foster efficiency and effectiveness, cultivate confidence, and promote a professional image.

HOW THE SYSTEM FITS WITH THE OTHER SYSTEMS

The forms presented in this chapter lay the groundwork for the initial steps in the sales process. They provide the basis for the administrative and sales staff to set up an ongoing multifaceted weekly and monthly monitoring and reporting system described in this chapter, which follows. Once the initial forms are completed, the builder's staff can proceed to assist purchasers with financing alternatives, selections, options and upgrades, and, finally, preparations for closing. It will become readily apparent that there is a synergistic relationship between many of the forms as information is collated and transmitted from one form to another as they are regularly revised and updated.

PROSPECT INFORMATION CARD

DESCRIPTION

The **Prospect Information Card** is a welcome and registration form used to obtain basic demographic information about the prospect and their housing requirements. We prefer a 5 × 7 NCR (no carbon required) card in triplicate. An 8½ × 11 sheet of paper, often consecutively numbered, is widely used but more unwieldy.

PROSPECT INFORMATION CARD

Welcome to <u>Builder's name and name of community</u>

Please take a moment to answer the questions below so that we may be better able to serve you.

Name:_____ Date:_____

Address:_____

City:_____ State:_____ Zip Code:_____

Home Phone:_____ Business Phone:_____

Cell Phone:_____ E-mail:_____

Employed by:_____

Present Resident	Number of children_____	Learned about us through:	Employment location:
❑ Rent Apartment/Home	Ages_____	❑ Times Herald Record	❑ Orange
❑ Own Home		❑ Journal News	❑ Rockland
❑ Living with Parents	Looking for a new home because:	❑ Signs/Drive by	❑ Westchester
	❑ Prefer to own vs. rent	❑ Harmon Homes	❑ New York City
	❑ Prefer site's location	❑ Newspaper ad in:	❑ Other_____
How many adults will	❑ Relocation	_____	Total annual income:
occupy your new home:	❑ Other_____	❑ Live in neighborhood	
Number of adults_____		❑ Our website	_____
		❑ The internet	
Ages of adults	Anticipated occupancy:	❑ Realtor	
❑ 20 to 29	❑ 3 months	❑ Other_____	
❑ 30 to 39	❑ 6 months		
❑ 40 to 50	❑ 1 year		
❑ 50 to 60	❑ Other		
❑ 60+			

(Backside of The Prospect Information Card)
PROSPECT RECORD

Rating and Comments:_____

Co-Broke Company:_____ Date of Intro:_____

Agent's Name:_____ Lic. No.:_____

Office Phone:_____ Home Phone:_____

Follow-up
Initial Reason to Call:_____

Return Visits: Date:_____ Comments:_____

Date:_____ Comments:_____

Date:_____ Comments:_____

Follow-up Records
Thank you Note:

Phone Call: Date:_____ Comments:_____

Direct Mail: Date:_____ Comments:_____

Phone Call: Date:_____ Comments:_____

Deleted: Date:_____ Comments:_____

Other Comments:_____

_____

_____

_____

PURPOSE

The sales counselor should review the prospect information card for completeness and legibility before the prospect leaves and immediately thereafter make notations while impressions are still fresh. Later on it's more difficult to recall the details of a specific visit, especially on a busy day. It may be discovered too late that the information is incomplete or illegible, rendering follow-up impossible.

The basic purpose of the prospect information card is to obtain information to build rapport so that the sales counselor can effectively assist the prospect in becoming a purchaser. The completed prospect information card discloses information about the prospect's current residence, their family make-up, the reason(s) they're looking for a new home, their time frame, their preferences for size and style, total family income (sometimes intentionally omitted), their work and employment location, how they learned about the community, and more.

By acknowledging the prospect's responses and perhaps relating elements of their situation to others in similar circumstances who have purchased new homes in the community, a cordial and empathetic relationship can be initiated. The sales counselor can then begin to qualify the prospect and tailor his or her presentation and the demonstration of homes accordingly.

The reverse side of the prospect information card is used to rate the prospect's purchasing potential, note comments or questions that require follow-up, and log future contacts and their outcome.

Because few prospects make a purchase decision on their initial visit to a community, it is imperative that a prospect information card be completed for every visitor to allow for follow-up. Unfortunately, our visits to sites over the years have confirmed the fact that prospect registration is irregular at best and too often totally neglected! This is unacceptable considering the investment made in their community, the cost of attracting prospects to a site, and the time they spend there.

USING THIS FORM

Who Prepares It and Who Updates It

A marketing or creative media firm often prepares the prospect information card with guidance from the sales staff. Its front side is rarely revised, but the reverse side is updated by the sales staff to note future contacts and return visits.

Who Uses It

The sales staff should have every visitor to the community complete a prospect information card. As we shall see, the raw data gleaned from the individual prospect information cards and collated on the weekly traffic report will allow the sales staff and marketing agency to compare and evaluate the relative cost effectiveness of the various components of the marketing program.

How to Use It

The sales counselor should read the prospect information card and use the information as an icebreaker to initiate conversation and commence a comfortable relationship. After the prospect leaves, the sales counselor should rate the customer's buying potential according to a predetermined scale (e.g., A = ready, willing, able, and interested; B = willing but not ready or able; C = not ready, willing, or able; X = could not determine interest level, noncommittal) on the reverse side. Additional comments or questions that require a response should also be noted, as well as future communications and their outcomes and any return visits. The sales counselor should initial the prospect information card so that if the prospect returns, they can be brought together again. If the sales counselor is not available when the prospect returns, another sales associate can at least review the card and take up where their colleague left off.

A-rated prospects should have their basic information transferred to the top ten prospect list as a reminder for systematic follow-up until they "buy or die."

Once a community is sold out, the prospect information card file box may well provide an initial list of prospect leads for the builder's next community.

How to Organize It

Entries on the reverse side should be chronologically recorded. The prospect information cards should be alphabetically sorted in a prospect card file box that remains in the sales office at all times. As the cards accumulate over time, it will be necessary to acquire a second box and subdivide the cards A to M and N to Z. It may be useful to label and file the prospect information card boxes annually and start fresh each year or at least clean out and file those cards that are more than a year old.

Who Gets Copies

The hard copy should be filed in the prospect card file box. The sales counselor should retain a copy for follow-up (perhaps from home in the evening without fear of misplacing the original). Third copies might be compiled for computer data entry.

Method of Transmission

The prospect information cards are hand filed.

Where It's Filed

The hard copy should be filed in the prospect file card box. The sales counselor should retain a copy in his/her personal files. The third copy can be used for data entry that's stored in a computer for label making for future mailings.

COMMUNITY COMPARISON CHART

DESCRIPTION

The **Community Comparison Chart** may be either a standalone display or a brochure insert, as in the example that follows, or both.

PURPOSE

The community comparison chart is a sales aid to differentiate the community from its competition by highlighting its appealing, valuable, and unique features (often referred to as "ESPs" or "USPs," standing for Exceptional or Exclusive Selling Propositions or Unique Selling Propositions). It is similar in use to the charts resale agents give to their customers or clients for evaluating the positive and negative features of properties they've seen. The community comparison chart serves as a quasi-comparative market analysis form to facilitate a final purchase decision.

USING THIS FORM

Who Uses It

Sales counselors use it to highlight the advantages of their community while indirectly pointing out the shortcomings that may be characteristic of competitive communities. Sales counselors should be fully informed about the nature of competitive offerings and undertake periodic visits to update their information. The community shopping report and sales evaluation form discussed in Chapter 8 is ideal for this purpose. The community comparison chart and the shopping reports will prepare the sales counselor to ably handle the inevitable comparisons as prospects visit several communities, gradually eliminating some from consideration and re-visiting others to narrow down their final selection.

How to Use It

Sales counselors should use the community comparison chart to promote the benefits that residents will receive by choosing their community over others under consideration. For example, the community comparison

COMMUNITY COMPARISON CHART

Please look over this standard checklist and take it with you when you look at other homes and communities. The standards at Dalton Farm are truly unrivaled elsewhere.

DALTON FARM STANDARD APPOINTMENTS

Does the prospective community have:

	Dalton Farm Yes	No	Other Community Yes	No
1. Hundreds of acres of splendid, preserved open space?	X			
2. 6,000 square foot clubhouse with elegant, ivy covered walled gardens, broad bluestone terraces, library, 5 fireplaces and a charming stone teahouse?	X			
3. An Equestrian Center with 48 box stalls, indoor/outdoor rings, riding trails, 40 acres of paddocks, classes and shows?	X			
4. Miles of cross country skiing and walking trails?	X			
5. Tennis courts, pool and area for children's play and picnicking?	X			
6. Trout streams?	X			
7. Quiet serene nature observation points?	X			
8. Spring fed lily pond?	X			
9. Superior Arlington School District?	X			
10. Walking access to Town Recreation area?	X			
11. Low maintenance and taxes?	X			
12. State-of-the-art Water and Sewer Service housed in architecturally exciting structures?	X			
13. Wonderful neighbors?	X			

If you are looking at a Townhouse does it have:

	Dalton Farm Yes	No	Other Community Yes	No
14. Two car garages?	X			
15. Elegant tiled raised entry foyer?	X			
16. Private courtyard?	X			
17. Private decks?	X			
18. Windows on three and four sides?	X			
19. Your own front and back yards?	X			
20. Basements?	X			
21. Cathedral Ceilings?	X			
22. Expanses of glass?	X			
23. Fireplaces with marble surround?	X			
24. Dining room and eat-in kitchen?	X			
25. Air conditioning?	X			

	Dalton Farm Yes	No	Other Community Yes	No
26. Cable television outlets in Living room and master suite?	X			
27. Telephone outlets in kitchen and master suite?	X			
28. Numerous artifact/plant shelves?	X			
29. Art niches?	X			
30. Laundry with full size Heavy Duty GE washer and dryer?	X			
31. Powder room?	X			
32. An unattractive centralized holding area for garbage?		X		
33. Garbage recycling center located off kitchen in garage?	X			
34. Acres of ugly parking?		X		

If you are looking at a Single Family Home, does it have:

	Dalton Farm Yes	No	Other Community Yes	No
35. Fabulous views?	X			
36. Cozy wood burning fireplaces with marble surround?	X			
37. Large accent windows?	X			
38. Soaring ceilings?	X			
39. Spacious walk-in closets?	X			
40. Dining room and eat-in kitchen?	X			
41. Numerous plant/artifact shelves?	X			
42. Private deck?	X			
43. Two-car garage with recycling area?	X			
44. Basement space for recreation, office, guest room, storage or utility use?	X			
45. Cable outlets in Living room and Master bedroom?	X			
46. Telephone outlet in kitchen and Master bedroom?	X			

Does the kitchen of the Townhouse or Single Family include:

	Dalton Farm Yes	No	Other Community Yes	No
47. Breakfast alcove?	X			
48. Garden views?	X			
49. Built-in GE self-cleaning range, dishwasher and refrigerator?	X			
50. Space saving pantry?	X			

Does the Master bedroom suite of the Townhouse or Single Family include:

	Dalton Farm Yes	No	Other Community Yes	No
51. Vaulted ceilings?	X			
52. Spacious walk-in closets?	X			
53. Plant/artifact shelves?	X			
54. Magnificent master bath with cultured marble garden tub and cultured marble double vanity?	X			
55. Separate shower stall?	X			
56. Private water closet?	X			

Does Exterior detailing of the Townhouse or Single Family include:

	Dalton Farm Yes	No	Other Community Yes	No
57. Dramatic architectural design?	X			
58. Poured Concrete Foundations?	X			
59. Underground cable, electric and telephone?	X			
60. Waterproof exterior outlets?	X			
61. Decks that are designed as an extension of the living area?	X			
62. Planning and design that preserves the extraordinarily beautiful natural environment?	X			
63. Views and privacy created through sophisticated site design?	X			

Are comfort and energy ensured with:

	Dalton Farm Yes	No	Other Community Yes	No
64. Low maintenance double glazed vinyl windows?	X			
65. Energy efficient oil fired HVAC (TH), energy efficient oil fired baseboard heating (SF)?	X			
66. DuPont Tyvek total house seal?	X			
67. R19 exterior wall R30 ceiling insulation?	X			

Do the houses you are looking at have these choices:

	Dalton Farm Yes	No	Other Community Yes	No
68. Full line of options and upgrades?	X			
69. Custom packages?	X			

chart will note standard features that are optional extras or not even offered by the competition; options and upgrades not available elsewhere; onsite amenities to enhance their enjoyment; special incentives or attractive financing; the reputation of the local school system; proximity to hospitals, houses of worship, shopping, and transportation. Although the decisive factor(s) may differ for each prospective purchaser, a thoughtfully designed community comparison chart will predispose prospects to select one community over others.

We do not advocate speaking negatively about the competition. Let your killer community comparison chart "kill it" for them.

How to Organize It
The community comparison chart should be organized in a logical sequence of categories that comprise the features and benefits of the community. Categories might include recreational amenities; brand name products; exterior and interior design features with emphasis on specific rooms; options, upgrades, and custom packages; special financing or other incentives; warranties; and neighborhood ambiance and attractions. Blank spaces should be provided so that comparisons with other communities can be easily noted or checked off.

Who Gets Copies
Every prospective purchaser should be given a copy of the community comparison chart and encouraged to compare and contrast your community with any others under consideration.

Method of Transmission
The community comparison chart is usually a hand insert in the community brochure that is distributed in the sales office. It may also be mailed out in response to requests for information.

Where It's Filed
The community comparison chart might be a display in the sales office or filed with other brochure inserts such as the site map, floor plans, features list, price list, available inventory list, information about the builder, and other useful brochures and handout materials.

INVENTORY AVAILABILITY LIST

DESCRIPTION

The **Inventory Availability List** is a compilation of the home sites (or lots) or units (in the case of condominiums or co-operatives) and their special features that are available for sale at a given date. The inventory availability list may be combined with the price list and other financial information or it may stand alone. A combined list would include information on taxes, monthly fees, if any (e.g., HOA fees or common charges), water and sewer fees, initial membership, and subsequent annual membership fees, if applicable.

PURPOSE

The inventory availability list is designed to control the inventory and provide for an orderly pattern of sales in the construction of the community. The available home sites or units are listed together with their size (acreage or square foot measurement), special features (e.g., lake view, allows for walkout basement or side entry garage), and limitations, if any (e.g., only certain models can be built on the home site). Premiums are noted and explained (e.g., $15,000 for a lake view, $12,000 for an oversized lot, or $10,000 for an end [corner] unit). Delivery dates are also stated. The inventory availability list is, of course, also a sales aid used to assist prospects in their choice of sites for their new home.

LOT AVAILABILITY LIST (A)

The Townhomes

Five beautifully designed and spacious Townhome and Ranch models are presented.

All feature a 2-car garage, fireplace with marble surround, vaulted ceilings, complete with General Electric refrigerator, range, dishwasher, full size washer and dryer, oil-fired forced heat and air conditioning. Basements are available depending upon townhome location.

All townhome owners enjoy the community's incomparable amenity package.

MODEL NAME	SQ. FEET	DESCRIPTION	MO. HOA CHARGE	EST. MO. RE TAX
Amsterdam	1253	1 BR/Loft/1.5 Bth	$105.44	$215
Ithaca	1368	2 BR/2.5 Bths	$105.44	$230
Syracuse	1499	3 BR/2.5 Bths	$105.44	$252
Geneva	1640	2 BR/Loft/2.5 Bths	$105.44	$275
Rochester	1533	Ranch/2 BR/2 Bths	$105.44	$279

✲ ✲

Available Townhome Inventory as of _____

MODEL	BASE PRICE	**BSMT OPT Y / N	LOT #	BLDG.#	PROJ. DEL.
* Amsterdam	SOLD	Y	123	1	Dec. 15
Rochester	$196,600	Y	124	1	Dec. 15
* Geneva	$199,800	Y	82	9	Feb. 15
Amsterdam	SOLD	Y	81	9	Jan. 25
Ithaca	$162,350	Y	80	9	Jan. 31
Syracuse	SOLD	Y	79	9	Jan. 15
* Geneva	SOLD	Y	78	10	Mar. 15
Ithaca	$162,350	Y	77	10	Mar. 15
Syracuse	$177,500	Y	76	10	Mar. 15
Syracuse	$182,600	N	86	14	Apr. 15
Ithaca	$166,600	N	87	14	Apr. 15
* Amsterdam	SOLD	N	88	14	Apr. 15
Geneva	$199,800	N	89	14	Apr. 15

* Amsterdam - $153,900   Base Price
**Basements are available at an additional cost and where terrain allows.

List is subject to errors, omissions, and changes without prior notice.

INVENTORY AVAILABILITY LIST (B)

Date:_____

Lots with Binders

Lot #	Premium	Acreage	Special Notes
77	$5,000	.535	Any Home can have side entry garage
83	$19,500*	.521	Any Home can have side entry garage, Lake view
89	$0	.292	Victorian Only
92	$0	.276	Any Home

Available Lots

Lot #	Premium	Acreage	Special Notes
84	$17,500*	.562	Claridge Only, Lake view
85	$17,500*	.535	Any Home, Victorian Only—side entry garage, Lake view
86	$17,500*	.584	Any Home, Victorian Only—side entry garage, Lake view
87	$17,500*	.456	Dorchester Only, Lake view
93	$0	.293	Chatham II Only

Model & Spec Homes

Lot #			Special Notes
96			Claridge Model
98			Victorian Model

*Premium price includes a Walkout Basement, 10′ × 12′ deck, (1) Patio Door & (1) Window in Basement

Note: Delivery Date is one year from the signed Purchase Agreement date.

List is subject to errors, omissions, and changes without prior notice.

USING THIS FORM

Who Prepares It and Who Updates It

The builder and his or her construction or sales staff prepare the inventory availability list and generally review it weekly for revision as home sites or units go into reservation or contract and are removed from the list while new ones that become available are added. Changes in prices, premiums, taxes, fees, and delivery dates would also be updated. The inventory availability list should be dated and numbered to reflect successive revisions.

Who Uses It

The sales counselor uses the inventory availability list to guide prospects in their selection of a home site or unit for purchase. Of course, the sales staff should be familiar with each inventory offering and be able to identify its comparative advantages and justify any price premiums. Each home site or unit should be evaluated for its location, size, views, privacy, amenities, delivery schedule, proximity to roadway(s) or overhead power lines, and other features to determine its relative desirability and price. Based on the home sites building envelope and the setback requirements, certain lots may only allow for specific model types to be built. Often the builder may designate specific lots for a single style of home to vary the mix of residences or to ensure that only certain styles and size homes are grouped exclusively in a specific location in the community.

Prudent builders have learned to save their finest lots or units for the final phase or section of their communities. They devise pre-planned and built-in marketing strategies for increasing prices from pre-construction through each successive phase of construction.

Practiced sales counselors will often ask interested prospects who have not quite made up their mind to proceed to try to narrow down their site selection and then to pick an alternative or second choice "just in case this one is sold before you return." Fear of loss is a great motivator. The inventory availability list and the price list, whether single forms or combined, should have a disclaimer stating "SUBJECT TO ERRORS, OMISSIONS, AND CHANGE WITHOUT PRIOR NOTICE."

How to Use It

Once a prospect has determined the specific model or style of home or unit, the sales counselor uses the inventory availability list to assist the prospect in determining their preference from the available inventory. The ultimate choice may involve several considerations: size and style of home, base price, premium price, location in the community, amenities, delivery schedule. The lot or unit may be the "last of its kind" —that is, the last to feature a lake view or provide a walkout basement, the last able to accommodate a specific style or model type or be completed within a specific time frame (e.g., for the opening of the school year), or the last in a building phase before a price increase goes into effect. It's useful to indicate on the inventory availability list which home sites or units are currently in reservation or contract and not immediately remove them from the list. Doing so shows sales momentum and helps to create urgency to make a decision. The inventory availability list is used in conjunction with the community site map, which is often color coded with dots or pins or flags to note title passed, contracts, reservations, and "courtesy holds" (see Chapter 5).

How to Organize It

The inventory availability list may be arranged simply by listing the home site number, the size of the lot, premium, if any, and notes indicating special features or restrictions and delivery dates. Alternatively, it may or may not be combined with a price list but could state the model types, square footage, descriptions, and financial information such as estimated real estate taxes and HOA fees or common charges, if applicable. Standard features (e.g., two-car garages, central air, fireplace) might also be described as well as a few of the more popular options (basements or finished basements, hardwood floors, or bonus rooms) and their prices.

Who Gets Copies

The sales staff, the construction staff, and prospective purchasers should have copies.

Method of Transmission

The inventory availability list is an intra-office form transmitted by fax or hand delivery.

Where It's Filed
The inventory availability list should be filed in its own binder.

CO-BROKERAGE PROTOCOL

DESCRIPTION

The **Co-Brokerage Protocol** is an outline of the procedures to be followed to create a cooperative relationship between the community's sales staff and the realtor community. A most insightful and comprehensive discussion of the topic can be found in Dennis Radice's *Sales Management Tool Kit: Working with Brokers, Agents, and Onsite Associates,* a BuilderBooks publication of the National Association of Home Builders.

PURPOSE

The co-brokerage protocol is designed to detail how a builder can increase a community's sales by implementing an effective and mutually beneficial co-op program with the real estate community.

Who Prepares It and Who Updates It
The co-brokerage protocol is prepared and modified, if needed, by the builder and the sales staff.

Who Uses It
The sales staff uses the co-brokerage protocol with brokers and agents who have customers or clients that are potential purchasers in the community.

How to Use It
The sales staff should visit all of the real estate offices in the vicinity of the community and others who service its general market area to introduce themselves and their development. A copy of the co-brokerage protocol and the realtor prospect registration and confirmation form (which follows) should be reviewed with the principal broker or office administrator, or both, so that the community's policies and procedures are clearly set forth and mutual roles and expectations are defined. It's best to call in advance for an appointment and try to arrange for a presentation to the entire staff at its regular meeting. An ample supply of brochures and forms should be left at each office visit.

Often, a broker or agent will bring a prospect to the community in response to an advertisement that includes the words "brokers welcome" or "brokers protected," without having had prior contact with the site. When that occurs, the agent should complete the realtor prospect registration form and be given a copy of the co-brokerage protocol to review while the prospect is completing a prospect information card.

The co-brokerage protocol should be signed and dated by the sales counselor and a copy given to the agent.

How to Organize It
Dennis Radice's guidelines for designing an effective co-op program are worth repeating. The program should be in writing, competitive, consistently applied, easy to learn, use, and administer, and, finally, avoid penalizing the onsite sales counselors. It should cover registration procedures, brokerage fees, communications, contracts, closings, and dispute resolution. A well-conceived co-brokerage protocol will maximize the relationship and avoid the potential pitfalls and problems inherent in such arrangements.

The introduction of the co-brokerage protocol at realtors' offices is generally the first step in managing the builder-broker relationship as a profitable partnership. The community might follow up with a Brokers Open House, periodic seminars, special events, contests, and mailings. Realtors who successfully sell the community should receive public recognition and be rewarded with referrals from the site where appropriate.

CO-BROKERAGE PROTOCOL

Date:_____

The staff at *Builder's name and community's name* values your professionalism and cordially invites you to participate in a lucrative Co-Brokerage Program at our community. Our success depends on your active co-operation and willingness to work with us, and we look forward to a mutually rewarding, smooth working relationship. To help ensure that result and to prevent misunderstandings, we have implemented the following Protocols:

To earn a 2½% Co-Brokerage Fee, based on the home's Base Sales Price Only, excluding options and upgrades:

1. Always register your customer in our co-brokerage program by filling out the realtor prospect registration form on your prospect's first visit. This will also serve as your confirmation and protection.

2. Call our sales office at (123) 456–7899 and provide your name, your firm's name, address and telephone number along with the full name, address and telephone number of your prospective purchaser(s) together with the relevant qualifying information (needs, income, urgency, etc.) *PRIOR* to bringing your prospect to our community.

3. Our sales staff will then check our customer and broker files to ascertain whether or not the prospect has had prior contact with us or has been registered with our sales office. If not, we will gladly register your prospect for a 6-month period. (Registrations may be renewed in writing for an additional 3 months.)

4. We have no control over a prospect changing brokers. If that occurs, we will register both brokers. To earn a fee, the broker or broker associate or salesperson must be the procuring cause of the sale. If there is a dispute between brokers, we reserve the right to place the funds in an escrow account until the matter is resolved.

5. Always accompany your prospect on his initial visit to the sales office and stay in contact with us, providing feedback. We reserve the right to follow up directly with any prospect brought to our community. Once a sale is made, our sales staff will follow through on the necessary details and steps to ensure the closing, and we will keep you informed of the progress and of any problems that may require your assistance.

6. Shortly before the closing, we will ask you to submit a brokerage fee bill. Fees are paid at closing. You are invited to attend but your presence is not mandatory.

Please have your principal broker or office manager sign this document and return it to us at your earliest convenience. Our sales office is open daily 11 AM to 5 PM, earlier or later by special appointment. We cannot accept registrations left on our answering machine and we ask that you provide ample time for us to give your prospect a complete presentation.

_____

Sales

_____

Principal Broker/Office Administrator

_____

Realtor Office

Who Gets Copies
Every real estate office that services the local marketplace should receive a copy of the co-brokerage protocol as should every broker or agent who visits the community. A copy should be filed in the site's co-brokerage protocol binder.

Method of Transmission
It's most effective to present the co-brokerage protocol in person at realtors' offices. The form also should be distributed at the sales office.

Where It's Filed
Co-brokerage protocols should be filed together in their own binder. If the prospect executes a purchase agreement, a copy of the co-brokerage protocol and the realtor prospect registration and confirmation form should be transferred to the prospect's file folder.

REALTOR PROSPECT REGISTRATION AND CONFIRMATION FORM

DESCRIPTION

The **Realtor Prospect Registration and Confirmation Form** serves as a written acknowledgement that a realtor, broker, or agent has registered a customer or client with the community's sales staff.

PURPOSE

This form activates and implements the co-brokerage protocol for a specific sales transaction. The purpose of the realtor prospect registration and confirmation form is to provide assurance and corroboration to an agent that is recognized by the sales staff as the party who has introduced a prospect to the site and who would be entitled to a stipulated brokerage fee if their prospect consummates a sale. It protects the agent as the procuring cause of the sale and confirms his/her right to a fee at the closing.

USING THIS FORM

Who Prepares It and Who Updates It
The builder and the sales staff prepare the realtor prospect registration and confirmation form and revise it as necessary.

Who Uses It
The sales staff uses the realtor prospect registration and confirmation form with outside agents who bring prospects to the community.

How to Organize It
The realtor prospect registration and confirmation form provides for detailed demographic information about the prospect and the registering agent and their real estate office. It reiterates the brokerage fee stated in the co-brokerage protocol and the duration of the validity of the registration and how it can be extended. It calls for mutual communication concerning the prospect's purchase decision and progress if they go forward, thanks the agent for his/her participation, and is signed by the agent and by the community's sales representative.

Who Gets Copies
The agent receives a copy for his/her office and personal records and the sales counselor retains a copy for the community's records.

REALTOR PROSPECT REGISTRATION & CONFIRMATION

Agent/Broker _____

Agency Type: Seller_____ Buyer_____

Street Address _____

City, State, Zip _____

Telephone # _____

Prospect Name _____

Street Address _____

City, State, Zip _____

Home Telephone _____ Work Telephone_____

Date of Initial Visit _____

_____ agrees to pay the agent/agency above a commission of two and one half percent (2.5%) of the base sales price on the sale of the home, exclusive of options and upgrades, when, if, and as title passes, providing the prospect has not previously registered at _____.

This brokerage fee policy is valid only when countersigned by seller's agent. This registration is valid for _____ days and may be updated by written notice for an additional _____ day period.

Thank you for registering _____ at our sales office. We look forward to working with you and your prospects over the coming weeks. Please call us if you have any questions about our community or procedure and provide us with feedback as well.

Agent/Broker Signature_____ Date_____

Representative_____ Date_____

Method of Transmission

Agents are given their copy at the sales office. If for some reason this does not occur, then their copy should be faxed or mailed to their office.

Where It's Filed

The sales counselor should attach the realtor prospect registration and confirmation form to the prospect information card, which should also have the agent's business card attached to it. If the prospect becomes a purchaser, the prospect information card with the realtor prospect registration and confirmation form is transferred to the purchaser's file folder.

To stand out from the competition, a builder might stipulate in the co-brokerage protocol and realtor prospect registration and confirmation form that they will pay half of the brokerage fee on contract signing and mortgage approval and the balance at closing. If the transaction fails to close title, any advanced fees will be promptly reimbursed, on written notice. This will eliminate the long waiting period that agents often encounter while waiting for a closing to collect their fees and thereby encourage their participation in the community's sales program.

RESERVATION FORM

DESCRIPTION

The **Reservation Form,** sometimes referred to as a binder or deposit, is an outline of the terms and conditions of the sale of property, which effectively removes a home site or unit from the available inventory for a stipulated time to allow for the execution of a purchase agreement or contract of sale.

PURPOSE

The reservation form details the agreement between the purchaser and the builder (or sponsor of a co-operative, condominium, or homeowner association unit) so that they can formalize their intentions in a purchase agreement. It is evidenced by the signatures of the parties and a good faith deposit that will become part of the purchaser's downpayment. The builder will have an attorney or the staff prepare the purchase agreement (usually in quadruplicate copy) and forward it to the purchaser's attorney.

USING THIS FORM

Who Prepares It and Who Updates It

The builder's attorney prepares the reservation form and revises it as necessary.

Who Uses It

The sales counselor has the purchaser complete the reservation form as evidence of his/her desire and intention to purchase the property. Its information is then used to prepare the purchase agreement and is transferred onto the memorandum of agreement.

How to Use It

The sales counselor should be certain that the prospect is qualified to proceed before completing the reservation form and certainly before the purchase agreement is mailed to the attorney. To facilitate the process, the sales staff should be able to qualify purchasers or have one or more mortgage lenders service the site so that onsite financing and qualification is readily accessible. A list of recommended local attorneys and lenders also should be available for distribution.

We recommend that a "back-up" or secondary reservation be taken whenever there is doubt that the primary purchaser is proceeding in a timely manner.

Occasionally, a purchaser may request a copy of the reservation form for their attorney to review prior to signing it.

RESERVATION FORM

Builder's name and name of community
Homeowners Association
(Nonbinding Reservation, after Receipt of the
Offering Plan and all filed Amendments)
Subject to Review and Acceptance by the Builder/Sponsor

On this date:_____   I, We_____

Reserve Lot#_____   Model_____

and agree to reserve said property under the following terms:

A)   Base Price   A) _____
B)   Lot Premium   B) _____
C)   Total Sales Price   C) _____
D)   Deposit (Binder)   D) _____
E)   Payment on Signing of Seller's Contract   E) _____
F)   Mortgage To Be Obtained   F) _____
G)   Cash Balance At Closing   G) _____

Contracts to be drawn and presented on or about_____
Transfer to Title to be completed on or about _____
Other Agreements_____
Reservation Deposit of $_____, made payable to: *"Attorney's Name Escrow Agent"* to expire
10 days from the above date and can be extended at the discretion of the Seller. (Reservation deposit is fully
refundable.)

Buyer(s) Information:
Name(s):_____
Address: _____
Telephone Numbers: His Home:_____   Her Home:_____
                         His Work:_____   Her Work: _____
                         His Income:_____   Her Income:_____
                         His Soc. Sec.#:_____   Her Soc. Sec.#:_____

The buyer(s) state that he/she _____(is or is not) represented by a Broker, other
than *Name of the On Site Agent,* in any connection with *Builder's name and name of community*

Broker Name and Office:_____

                                            _____
                                            Purchaser

_____   _____
Sales Representative   Purchaser

Attorney Information:
Buyer's Attorney:_____
Address:_____
Telephone:_____Fax:_____

RESERVATION FORM (CON'T)

In consideration of said reservation deposit, the sponsor hereby reserves for a period of ten (10) days after date hereof ("reservation period") the lot/unit for sale.

The refundable reservation/deposit will entitle the prospective purchaser to inspect the offering plan and amendments, if any, and to make selections within 10 days.

After the selections are completed, a purchase agreement will be drawn up and forwarded to the purchaser(s) or their attorney. Within 10 days of receipt of the purchase agreement, the prospective purchaser(s) will be required to return to the sponsor a signed purchase agreement with the downpayment (usually 10%) of the total purchase price.

Seven (7) days thereafter, the purchaser(s) will receive a fully executed purchase agreement. RECEIPT OF ALL FUNDS, EXCEPT THOSE DUE AT CLOSING, ARE SUBJECT TO COLLECTION AND SHALL BE HELD IN TRUST IN ACCORDANCE WITH SECTION 352h AND SECTION 352e2 (b) OF THE GENERAL BUSINESS LAW.

The home reservation/deposit shall be refunded in full, without interest, at any time before the purchaser(s) execute(s) a binding purchase agreement if purchaser(s) shall demand the return thereof, or without demand, if (and when) the sponsor executes a purchase agreement to sell the home to a third party after the expiration of the reservation period, but in no event later than thirty (30) days after the date thereof.

Purchaser(s) acknowledge(s) receipt of a complete copy of the offering plan and all filed amendments for the homeowners association. This nonbinding home reservation/deposit is not assignable.

Pursuant to the requirements of the Department of Law of the State of New York, a purchase agreement for the sale and purchase of a home may not be executed and exchanged between a sponsor and a prospective purchaser until after the prospective purchaser has had not less than 72 hours to review the homeowner association documents.

_____

Purchaser

_____

Purchaser

_____

Sales Counselor
_____

How to Organize It

The reservation form includes detailed purchaser information, the terms and conditions of the sale, the date by which the purchase agreement is to be executed, and the date for the title transfer or closing. It also includes the buyer's attorney information and information about the realtor, if applicable. Of course, it is dated and signed by the purchaser(s) and the sales counselor.

Who Gets Copies

The purchaser, their attorney, and the seller's attorney receive copies of the reservation form.

Method of Transmission

The reservation form is included with the memorandum of agreement and purchase agreement sent to the respective attorneys.

Where It's Filed

The sales counselor affixes a copy of the reservation form in the purchaser's file folder.

CANCELLATION FORM

DESCRIPTION

The **Cancellation Form** states the reason(s) why a sale falls through or is canceled before the transfer of title or settlement takes place. This usually occurs after a reservation is completed but prior to the execution of a purchase agreement. More rarely, a sale may be canceled after a contract of sale is signed but before the closing.

PURPOSE

The cancellation form is used to record the reason(s) the purchaser is unwilling or unable to proceed with the transaction and to determine what, if anything, the sales counselor or builder can do to salvage the sale. An excessive number of cancellations calls for a thorough review and examination of the causes and remedial action. The "fall through" rate varies from community to community, but there is usually a greater incidence of such occurrences on the more affordable sites because they attract a greater number of more marginally qualified prospects.

USING THIS FORM

Who Prepares It and Who Updates It
The sales counselor prepares the cancellation form and updates it if necessary.

Who Uses It
The sales counselor uses the cancellation form when the purchaser declares his/her intention to terminate the sale.

Perhaps the purchaser cannot come up with the full downpayment at one time. The builder may allow for a staggered downpayment schedule to save the sale.

How to Organize It
The cancellation form provides a listing of the most common reasons purchasers back out of transactions and prompts the sales counselor to consider what might be done to save the sale.

Who Gets Copies
The purchaser should be given a copy of the cancellation form along with the refund of his/her reservation deposit and/or contract downpayment money, if warranted. The now former purchaser should be asked to sign the cancellation form and a receipt for any funds that are returned to him/her. Before a refund is issued, however, the offering plan or prospectus, if applicable, should be returned to the sales office. Reservation and contract cancellations are noted on the weekly traffic report and are reflected on the monthly transactions (sales activity) report.

Method of Transmission
The sales counselor inserts the cancellation forms into a cancellation form binder.

Where It's Filed
The cancellation forms are filed in chronological or numerical order in their own cancellation form binder.

CANCELLATION FORM

Community:_____   Homesite# / Unit#:_____

Name of Purchaser:_____

Reservation Date:_____   Cancellation Date:_____

Reason(s) for Cancellation:

1. Cannot Afford:
    A. Cannot raise downpayment_____
        How much can be raised? $_____
    B. Cannot get gift letter_____
    C. Cannot get co-signer_____
    D. Cannot get financing_____
        Reason(s):_____
        _____
        _____
    E. Cannot afford monthly payments
        What is Mo. Income? _____
        Debts? _____
        Job Security?_____

2. Residence too small?_____
    What do they want?_____

    Residence too large?_____
    What do they want?_____

3. Do not like area _____
    What area(s) do they prefer?
    _____

4. Buying somewhere else?
    Where? _____
    Why?_____

5. Home sale fall through?_____

6. Particular objection(s):
    _____
    _____
    _____

7. What, if anything, could be
    done to make this deal?
    _____
    _____
    _____

Note: Try to have the applicant return in person so you can discuss alternatives and try to re-covert to a sale. Do not refund a deposit check where a prospectus or offering plan is involved (cooperative, condominium, townhome, HOA) until you receive the offering plan back in good condition.

MEMORANDUM OF AGREEMENT

DESCRIPTION

The **Memorandum of Agreement** is a summary of the basic facts of a sales transaction.

PURPOSE

The builder's or sponsor's attorney or sales director or administrative assistant will use the memorandum of agreement in conjunction with the reservation form to prepare the purchase agreement or contract of sale.

USING THIS FORM

Who Prepares It and Who Updates It
The sales counselor or administrative assistant prepares and revises the memorandum of agreement as needed.

Who Uses It
Whoever prepares the purchase agreement or needs a quick summary of the basic facts of a sales transaction references the memorandum of agreement.

How to Use It
Information from the reservation form is transposed to the memorandum of agreement. Any changes in the purchaser's address, phone number, or place of employment from the time of initial registration until closing is noted on the memorandum of agreement.

How to Organize It
The memorandum of agreement provides the basic information on the buyer, the seller, the attorneys, the buyer's broker or agent, the property address, purchase price, terms and conditions of the sale, and proposed closing date.

Who Gets Copies
The buyer's and seller's attorneys receive copies of the memorandum of agreement.

Method of Transmission
The memorandum of agreement is transmitted along with a copy of the reservation form and the purchase agreement to the attorneys by mail or hand delivery.

Where It's Filed
The memorandum of agreement is affixed to the purchaser's file folder.

PURCHASE AGREEMENT OR CONTRACT OF SALE

DESCRIPTION

The **Purchase Agreement*** is a legal document containing the terms and conditions for the conveyance of title to real property from a seller to a purchaser. It includes numerous stipulations defining the mutual re-

---

*This form varies based on state law, on the type of residence being offered, on common practice in the area, and, of course, on the law firm and the attorney preparing it. Each community should insert its own purchase agreement.

MEMORANDUM OF AGREEMENT

Date: _____

Seller: _____
_____
Address: _____
_____
Phone: _____ Fax: _____

Purchaser: _____
Address: _____
_____
Phone: _____ Fax: _____

Seller's Attorney: _____
Address: _____
_____
Phone: _____ Fax: _____

Purchaser's Attorney: _____
Address: _____
_____
Phone: _____ Fax: _____

Property Location: _____
_____
_____

Purchase Price: _____

Terms: _____
_____

Broker for Buyers: _____

Proposed Closing Date: _____

sponsibilities of the parties. The purchase agreement may also include a floor plan, a list of construction specifications, a list of standard features, a warrantee section, a W-9 Form (to allow for the downpayment money to be deposited in an escrow account), and a list of options and upgrades if they have been completed. If they have not been completed, the options and upgrades selections are appended later as a special exhibit, addendum, or rider.

PURPOSE

The purchase agreement is based on the preliminary reservation form and allows the sales transaction to proceed into a binding, contractual obligation. A lending institution will require a copy of the purchase agreement as one of the documents needed for mortgage financing. The builder's construction lender may require a copy of the purchase agreement to advance funds for building.

USING THIS FORM

Who Prepares It and Who Updates It

The builder's attorney prepares the purchase agreement from the material furnished in the reservation form and the memorandum of agreement. The attorney will revise the purchase agreement as circumstances warrant. Frequently, the sales director or an administrative assistant fills in the blank spaces in a prepared purchase agreement and transmits it.

Who Uses It

The purchase agreement will be used by the purchaser, the seller, their respective attorneys, the purchaser's lending institution, and the builder's construction lender.

How to Use It

The purchaser and the builder will use the purchase agreement to secure financing. The attorneys will reference it if a dispute or litigation arises between the buyer and seller. A sample purchase agreement is included in the offering plan or prospectus required in the sale of a co-operative, condominium, or homeowner association unit as part of the disclosure requirements. It is not unusual for a prospective purchaser to request to read or review all or part of the offering plan before deciding to proceed.

The sales staff should be completely conversant with the purchase agreement and closely monitor the stipulations concerning the time frame for mortgage application and commitment, selections, and other time-sensitive provision for compliance.

The sales staff should take a "back up" or secondary reservation at the slightest sign of a stalling tactic. If an original purchaser fails to perform in a timely manner, that purchaser risks not only timely delivery but also the loss of the property altogether, especially in a "hot" market where the next buyer may be more than willing to pay an even higher price.

Experience teaches that the longer a property is tied up by a purchaser's failure to execute the purchase agreement, the greater the likelihood that the "sale" will result in a cancellation. Delays may be the result of several factors: laxity in issuing the purchase agreement; the buyer's attorney "re-writes" the contract or appends a lengthy rider, or both; or the purchasers are stalling. Perhaps they are still looking at other properties or are waiting for mortgage approval on the part of the purchaser of their current residence or mortgage approval for their purchase, or for the availability of funds for the downpayment or until completion of their selections or their pricing. Depending on the reason a purchase agreement is not signed in a timely fashion, the builder may either grant an extension or exercise his prerogative to "kill the deal" by voiding the agreement and placing the property back on the market.

How to Organize It

The purchase agreement is organized by the seller's attorney in accordance with prevailing legal custom for contractual requirements based on the type of residence being conveyed. A typical purchase agreement with appended collateral materials may run 15 to 25 pages or longer of fine print.

Who Gets Copies
Purchase agreements are retained by the purchaser, the seller, their respective attorneys, the purchaser's lending institution, and the builder's construction lender.

Method of Transmission
The purchase agreement (usually in quadruplicate copy) is mailed or hand delivered to the parties involved. It is wise to provide a covering letter with the purchase agreement advising the seller's attorney of the site's procedures regarding changes or modifications.

Where It's Filed
The purchase agreement is filed in the purchaser's file folder.

ARCHITECTURAL AGREEMENT

DESCRIPTION

The **Architectural Agreement** is, in effect, a change order whereby with the builder's consent the purchaser may contact and contract with the builder's architect to draw custom changes to the elevation or floor plan of the home. The architect will prepare stamped sets of blueprints for the revised new plans for the builder to file with the local authorities and to distribute to the subcontractors and tradesmen.

PURPOSE

The architectural agreement allows the builder to provide the purchaser with the option of making custom changes, thereby helping to ensure the sale. It also has the benefit of having the purchaser work directly with the architect, eliminating any third-party misunderstandings.

USING THIS FORM

Who Prepares It and Who Updates It
The builder or the sales staff prepares the architectural agreement in conjunction with the architect. It is revised when necessary, usually to reflect a change in the fee structure.

How to Use It
The sales counselor and the purchaser will mark up the original plan to reflect the desired modifications and forward it along with clearly written instructions. The architect will then review the proposed changes and provide the purchaser with an estimate of the costs entailed. If they wish to proceed, based on the costs from the builder and the architect, the purchaser will sign the agreement and pay the architect directly. An initial nonrefundable deposit is required. The work customarily is not started until payment has cleared the bank.

How to Organize It
The architectural agreement is organized like a reservation agreement between the purchaser and the architect (seller). The fee structure and an estimate of the time and cost is stated. There is a signature line for the purchaser to indicate the acceptance and authorization for the architect to proceed.

Who Gets Copies
The architect, the purchaser, the construction department, and the sales office should receive copies.

ARCHITECTURAL AGREEMENT

Agreement made this _____ day of _____, _____, between

_____ ("Seller") and

_____("Purchasers").

    Our Architect,_____AIA, has agreed to provide design services for individual customers requesting plan changes.

    The fees for design services performed by _____ Architects for individual customer changes shall be based on the following hourly rates:

Principal	@$90.00/hr
Project Architect	@$70.00/hr
Draftsperson	@$40.00/hr

    If this Agreement meets with your approval, please sign below and return it with a nonrefundable $1,000.00 check made payable to "_____."

    No architectural work will commence until the check has been deposited and has cleared the account.

Accepted:

_____   _____
Purchaser   Date

_____   _____
Purchaser   Date

Method of Transmission

The architectural agreement may be transmitted by mail, hand delivery, or fax.

Where It's Filed

A copy of the architectural agreement is affixed to the selections section of the purchaser's file folder.

PURCHASER FAMILY PROFILE FORM

DESCRIPTION

The **Purchaser Family Profile Form** is designed to collect data about each family member, including place(s) of employment, finances and financing relevant to the purchase, and information about the search for a new residence.

PURPOSE

The purchaser family profile forms provide information that serves a variety of purposes. A composite picture of the typical purchaser family and vital statistics for each member can be drawn and used to refine and target market the advertising campaign to attract likely buyers. The builder will profit from learning why purchasers selected the community and chose a particular home style or model type. This information is useful in designing future offerings. The sales staff can use the purchaser family profile form data to inform prospective purchasers about other families and their children who may become the new friends and playmates of their children. The information also can be used to create a homeowners directory that will be useful to families, the sales and construction departments, and others. Finally, the new residents feel that the builder and their staff really care enough about them to solicit their opinions.

USING THIS FORM

Who Prepares It and Who Updates It

The sales staff prepares the purchaser family profile form and updates it when appropriate.

Who Uses It

The advertising or media agency can use the information in designing the advertising and selecting its placement. The builder can use the research portion in determining the design of future homes. The sales staff can relate useful information about existing purchasers to prospective purchasers. The financial information may be of use to a lending institution in offering tailor-made products to typical buyers.

How To Use It

The sales counselor or administrative assistant should have each family complete their profile form after the purchase agreement is executed and selections have been completed but prior to closing. It should be reviewed to search out similarities and patterns among the purchaser families, which can assist the sales staff in their presentations and provide guidelines and insight for advertising and construction decisions. The information also may be used to compose a homeowners directory for residents.

Although the prospect information cards can also be reviewed to search for patterns and similarities, a significant difference may exist between those who visit a community and its actual buyers–for example, do a significant number of purchasers come from the same zip code? Use a particular realtor? Have two children?

How to Organize It

The purchaser family profile form should have sections on basic demographics, employment, and financial data as well as questions pertaining to the

PURCHASER FAMILY PROFILE FORM

Date _____Community_____Reservation Date_____Contract Date_____

Model_____Lot/Unit#_____Street_____

Buyer No. 1 Name_____ SS#_____Age_____

Buyer No. 1 Name_____ SS#_____Age_____

Current Address_____

City/State/Zip Code_____Home Phone_____

Children: (1) Name and Age_____ (2) Name and Age _____

                (3) Name and Age_____ (4) Name and Age _____

                (5) Name and Age_____

EMPLOYMENT	Buyer No. 1	Buyer No. 2
Present Position:	_____	_____
Local Employer:	_____	_____
Employer's Local Address:	_____	_____
City, State, Zip Code:	_____	_____
Business Phone No.:	_____	_____
Length of Service:	_____	_____
Are You Self-Employed?	Yes_____No_____	Yes_____No_____
Annual Salary:	_____	_____
Overtime/Bonus:	_____	_____
(if continuous)	_____	_____
Other Income (specify)	_____	_____
Alimony/Child Support	_____	_____
Received:	_____	_____
Duration of Payments Remaining:	_____	_____
Total Annual Income:	$_____	$_____

Note: Alimony, child support, or separate maintenance income does not have to be revealed if the Buyer does not choose to have it considered as a basis for repayment of the loan.

BUYER NO. 1 Prior Employment (If Local Employment is less than 3 years)

Employer:_____ Annual Income:_____

Employment Dates:_____to_____

BUYER NO. 2 Prior Employment (If Local Employment is less than 3 years)

Employer:_____ Annual Income:_____

Employment Dates:_____to_____

FINANCIAL AND MORTGAGE INFORMATION

Rent or Own Present Residence?   Own_____ No. of Years_____ Rent_____ No. of Years _____

Name and Address of Lending Institution_____

_____

Name of Mortgage Officer_____ Phone No._____

Name and Address of Attorney for the Lending Institution _____

_____ Phone No._____

Name and Address of Title Company_____

_____ Phone No._____

Name and Address of Purchaser's Attorney_____

_____ Phone No._____

Purchase Price: $_____

Options and Upgrades: $_____

Credits: $_____

Subtotal: $_____

Initial Deposit: $_____

Subsequent Deposits Total: $_____

Mortgage $_____

Balance due at Closing: $_____

1st Mortgage: $_____ Interest Rate:_____Term:_____

Project Closing Date: _____

Pre-Closing Walk Through
   Inspection Date: _____

Actual Closing Date: _____

HOUSING RESEARCH

How long were you looking for a home?_____Number of homes inspected:_____

Number of communities visited:_____Did you work with a Realtor? Yes_____ No____.___

Please identify the major reason(s) you selected the home that you did and this community:

_____

_____

_____

Please indicate any special interests or hobbies of family members:_____

_____

_____

Additional comments:_____

_____

_____

_____

_____

search for new housing and reasons for the final selection. Sections should be included for any areas of additional interest and blank space provided to record additional comments. Purchaser family profile forms often request an evaluation of the sales staff and the construction staff, inquire about any areas where improvements can be made, and ask for referrals.

Who Gets Copies

The purchaser family profile forms should not be distributed because they include a great deal of personal information of a confidential nature.

Method of Transmission

The blank purchaser family profile forms should be delivered in person. When completed and returned, they should remain in the sales office.

Where It's Filed

The purchaser family profile form should be filed either in the purchaser's file folder or collectively in their own alphabetically arranged binder, or in both places.

CHAPTER 2

SETTING UP A MONITORING AND REPORTING SYSTEM

OVERVIEW

The forms presented in Chapter 1 were designed to be used with individual prospects. The forms introduced in this chapter are used to track and document the progress of multiple transactions once purchase agreements are executed. These forms survey the status of the community's sales program on a weekly basis except for the monthly revised transactions (sales activity) report and the summary annual report. These forms comprise an integrated monitoring and reporting system.

DESCRIPTION AND EXPLANATION

A vital community has a multiplicity of activities occurring simultaneously. These activities include servicing daily traffic and callers, taking reservations, contract-related work, preparations for closings, office maintenance and meetings, selections including options, upgrades, and change orders, mortgage-related work, filing and reports, and more. Various forms are required to maintain orderliness and to trace the flow of sales from initial community visits to final settlements. Forms provide the raw data for both short- and long-term analysis and help to pinpoint areas that require attention and improvement. As in the first chapter, the forms will be presented in the sequence in which they are most likely to be used. Each one will be described with its purpose and use explained, including who prepares it, who updates it, who gets copies, how it is transmitted, and who files it and where. A chart conveniently summarizes this material (see Forms list).

HOW THE SYSTEM FITS WITH THE OTHER SYSTEMS

The forms presented in this chapter logically succeed those presented in the previous chapter. They are used to record the progressive steps required in a successful sales transaction. A prospect's first contact with the community is recorded on the day of the visit on the weekly traffic report. The prospect's closing will also be noted on that report as well as on the monthly transactions report, on the comparable sales report, and on the annual report. Interim steps will be noted on the schedule of future closings report and on the preliminary checklist for closings. In the following chapters, a protocol for selections, options, and upgrades and forms used for mortgage financing will be separately considered. The interrelationships between the forms will become readily apparent as information is collated, revised, and transmitted from one to another when they are regularly updated.

WEEKLY TRAFFIC AND SALES REPORT

DESCRIPTION

The three-part **Weekly Traffic and Sales Report** provides for a daily accounting of the community's traffic and sales activity including the number of phone inquiries, the number of visitors, and how they learned about the site. The second section is used to list the reservations taken or the cancellations that occurred during the week. The third section is a summary of the week's overall activity, providing the number of phone calls, visitor traffic, reservations, cancellations, contracts, and closings. Space is also allotted to note advertising placements and relevant comments (e.g., introduced a new model, snowed on Saturday, Broker's Open House on Tuesday).

PURPOSE

The weekly traffic and sales report is used to monitor the community's traffic and sales activity on a daily basis with weekly summations. It is used to determine the relative strength and cost effectiveness of the various components of the marketing program and the relative success of the sales effort.

USING THIS FORM

Who Prepares It and Who Updates It

A new consecutively numbered weekly traffic and sales report should be prepared by the sales staff each week. The sales staff updates the report daily at the close of business and adds the daily numbers together at the end of the week, resulting in the weekly tally and the year-to-date totals.

How to Use It

The weekly traffic and sales report is used to provide information to evaluate the effectiveness of the marketing effort. The relative strength of the various components in the marketing mix to draw prospects to the community can be individually determined to learn what is working well and what is not. For example: Are the ads attracting prospects to the site? Are the flyers and mailers effective? Are the Internet and website significant factors? Is the co-brokerage program working? Are satisfied homeowners a source of referrals? Are billboards and other signs accounting for traffic? Based on the relative effectiveness of the various components, resources should be reallocated to enhance traffic and the likelihood of sales.

The weekly traffic and sales report data is also used to calculate and compare the cost effectiveness of the various components of the marketing mix as well by simply dividing the expenditure for a specific component by the number of visitors it elicited. This yields the cost per visitor for the time period measured. For example, if $12,000 was expended on newspaper ads that attracted 24 visitors over two weeks, the cost to attract a single visitor was $500 for that period ($12,000 ÷ 24 = $500). In our experience, realtor participation typically produces 15 to 33 percent of the total sales, and in some instances even more. Referrals from satisfied purchasers generally account for 5 to 15 percent or more of the total sales. Realtor and referral sales cut down on the marketing expenditures, generate goodwill, and shorten the sellout period.

Increasingly, house hunters are turning to the Internet and community websites in the search for new homes. Many multiple listing services and local builder associations automatically place their listings on the Internet, which is conveniently available 24 hours a day.

If expenditures are carefully monitored, the cost of each sale can be accurately determined. One should be careful, however, to distinguish clearly between the sources that generate traffic and those that account for "sales." There may be a significant difference rather than the expected direct correlation.

Perhaps the most revealing number of the weekly traffic and sales report is that of "return visitors" or "be-backs." Because most purchasers visit a site several times before making a final decision, the number of returns is a vital statistic. Experience has shown that a "healthy" site shows a return rate of 8 to 15 percent or better. That means that 1 in every 12.5 visitors to 1 in every 6.67 visitors actually purchases a home. Above we saw that typically only 3 to 5 out of 100 typically do so. The number of returns correlates directly with the number of sales. If a site does not measure up in this category, it's a certain indication of distress.

WEEKLY TRAFFIC & SALES REPORT #_____

Week of Monday_____   Prepared by_____

MONTH/DATE _____

Source	Monday	Tuesday	Wednesday	Thursday	Friday	Saturday	Sunday	Weekly Total	Y-T-D Total
Local papers									
NY Times									
Gannett									
Harmon Homes									
Pennysaver									
Other Magazines									
Referral-Nonbuyer									
Referral-Purchaser									
Seminar									
Mailers									
Signs									
Radio									
TV									
Brokers									
Website/ Internet									
Return Visitors									
Total Traffic									
Phone Calls									

Detailed Weekly Deposit Activity

Unit	Type	Bldg/Clstr	Purchaser(s)	Date Resv'd or Fall Thru	Scheduled Closing	Source	Sales Associate

WEEKLY SUMMARY		COMMENTS	
Traffic		Advertising	
Calls			
Binders			
Cancelled Binders			
Binders Outstanding By End of Week		Other Comments	
Contracts Sent Out This Week			
Contracts Signed This Week			
Cancelled Contracts This Week			
Total Contracts By End of Week			
Total Closed By End of Week			

The weekly traffic and sales report is also used to illuminate the sales effort. Experience has shown that only three to five percent of visitors to a community complete a reservation, which successfully converts to a contract and closing. We can calculate the closing ratios of the sales staff by comparing the total number of reservations, contracts, or closings with the total number of visitor prospect information cards. If it's determined that the number of closings is below industry standards or that there is an unusually high number of cancellations of reservations or contracts ("fall throughs"), the reason(s) must be ascertained and remedial action taken. Alternatively, if sales are briskly occurring, then pricing and procedures should be adjusted.

The weekly traffic and sales report should be carefully reviewed from week to week and the appropriate reallocation of marketing resources or other remedial measures undertaken as warranted. If viewed with a critical eye and probing mind, the weekly traffic and sales report will provide a gold mine of information that can be used to improve the bottom line.

How to Organize It

The tri-section organization is recommended. (If more reservations or cancellations are recorded in a week than the space allows for, use a second weekly traffic and sales report sheet.)

Who Gets Copies

The sales staff, the builder, and the construction department should receive a copy of the weekly traffic and sales report.

Method of Transmission

The weekly traffic and sales report should be hand delivered, transmitted by fax, by mail, or by e-mail.

Where It's Filed

The sales staff should file the weekly traffic and sales report in its own binder and keep it in a central location together with the binders for the other forms.

TOP TEN PROSPECTS LIST AND FOLLOW-UP FORM

DESCRIPTION

The **Top Ten Prospects List and Follow-Up Form** is a weekly updated log of the contacts and the results of actions taken with the most promising prospective purchasers.

PURPOSE

The top ten prospects list and follow-up form serves as a reminder to schedule immediate and continuous contact with ready, willing, able, and interested prospects until a purchase decision is made. Failure to complete a prospect information card is the number one deficiency of most sales associates, and failure to follow-up with prospects is a close second.

USING THIS FORM

Who Prepares It and Who Updates It

The sales staff should revise this form as circumstances dictate by adding new prospects and eliminating older ones (who may have decided to complete a reservation or who indicated they are no longer interested) and recording all communications (e.g., phone calls, mailings, e-mails) and the results.

TOP TEN PROSPECTS LIST & FOLLOW-UP REPORT

Date:_____

	Prospect	Phone #	Action	Response
1.	_____	_____	_____	_____
	_____			
2.	_____	_____	_____	_____
	_____			
3.	_____	_____	_____	_____
	_____			
4.	_____	_____	_____	_____
	_____			
5.	_____	_____	_____	_____
	_____			
6.	_____	_____	_____	_____
	_____			
7.	_____	_____	_____	_____
	_____			
8.	_____	_____	_____	_____
	_____			
9.	_____	_____	_____	_____
	_____			
10.	_____	_____	_____	_____
	_____			

How to Use It

Each week the top ten list should be reviewed to determine the most appropriate action to be taken with each prospect. Here are some of the many cogent reasons to contact prospects. Reasons include:

1. You have the answers to specific questions or the requested information.
2. A new section of homes is opening up for sales/ or a new model or style is being introduced.
3. A new builder incentive or financing program is being introduced.
4. The model merchandising has been completed.
5. New community information is now available (school report, shopping, amenities, medical center, sports stadium).
6. The prospect is invited to a party! (Barbecue, Golf, Pool, Clubhouse, Memorial Day, 4th of July, Holiday Season.)
7. The home that the prospect was interested in has come back on the market.
8. Your "courtesy hold" on Lot #77 is about to expire.
9. Only one home left of the model you preferred, or that can meet your move in date requirements or at the "old" price before the scheduled increase goes into effect.
10. The building specifications have been changed to enhance value.

At any one time, you may not have ten top prospects but if you have considerably more, you may not be closing effectively.

How to Organize It

The top ten prospects list and follow-up form can be arranged as a numerical list with the prospect's name, address, phone number, and e-mail address (taken from the prospect information card) with space to list the dates of communications and contacts and their outcome.

Who Gets Copies

The sales staff might find it beneficial to review this form with the builder or the construction super to decide the nature of the next contact and who is best suited to make it.

Method of Transmission

The top ten prospects list and follow-up form should be hand delivered because it is best discussed in a face-to-face meeting.

Where It's Filed

The top ten prospects list and follow-up form should be filed in its own binder.

WEEKLY TRACKING REPORT

DESCRIPTION

The **Weekly Tracking Report** provides a detailed chronological record of the status of each lot or unit sale in terms of its reservation, purchase agreement, mortgage application and commitment, selections, scheduled closings, and other pertinent information.

PURPOSE

The weekly tracking report monitors the progress of every sale from the reservation to the closing and provides for significant comments and key party contact information. Its numerical columns (base price, options and upgrades, credits, and total price) record the total expenditure for each individual sale and for the total of all sales. This single form supplies a summation of the community's sales activity at a glance, listing the

WEEKLY TRACKING REPORT

Project Name: _____

Date: _____

Lot #	Style	Buyer		Binder	Base	Lot	Options	Inc/prog	Tot Purch	Contract		Attorney		Mortgage				Closing	Comments
		Name	Telephone	Date	Price	Premium	Upgrades	Credits	Price	Mailed	Signed	Name	Phone	Lender	Appl Date	Comm Date	Contract	Sched.	
Totals																			

critical data for all of the sold (reservations and contracts) but not yet closed inventory. In effect, the weekly tracking report is a chronology of the unfolding events in the life cycle of each sale and a financial accounting of the sales to date.

USING THIS FORM

Who Prepares It and Who Updates It
The sales staff prepares and revises this form on a weekly basis (making daily notations as new information becomes available by filling in the blanks in the appropriate columns).

How to Use It
This form is used in tandem with the weekly traffic and sales report. As each new reservation is completed, the sale is noted on the weekly traffic and sales report and added to the next available line of the weekly tracking report. As each step in the sale process is completed, its date is entered under the appropriate heading until all twenty columns are completed. Of course, when a purchaser closes title, their line is deleted from this form. The closing is then recorded on the weekly traffic report and added to the comparable sales report.

The blanks on each line indicate which steps remain to be completed and pinpoint those areas that require attention and follow-up by the sales staff. For example: Is there an attorney for a specific purchase? Have contracts been sent out? Have they been returned? Has a mortgage application been made? With whom? Has a commitment been issued? Have selections been completed? Are there any problems with the transaction? In short, is the sale progressing in an orderly and timely fashion?

How To Organize It
The weekly tracking report is organized as a spreadsheet. An Excel program serves this purpose well.

Who Gets Copies
The sales staff, the builder, and the construction department should receive copies of the weekly tracking report together with the distribution of the weekly traffic and sales report.

Method of Transmission
The weekly tracking report should be delivered by hand or by fax.

Where It's Filed
The weekly tracking report should be filed in its own binder.

MONTHLY TRANSACTIONS (SALES ACTIVITY) REPORT

DESCRIPTION

The **Monthly Transactions Report** is a compendium of the community's reservations, purchase agreements, closed sales, and available or unsold inventory.

PURPOSE

The tri-section monthly transactions report provides a concise summary of the site's sales activity on a month-to-month, year-to-date, and overall basis from the inception of the sales program to the most recent month's activity. The lower and center sections serve as a progress report of the sales effort viewed on a monthly, bi-monthly, quarterly, semi-annually, or annual basis as the monthly updates are successively collated. This enables the entire staff to see if goals are being met.

MONTHLY TRANSACTIONS (SALES ACTIVITY) REPORT

TO: DATE: 2001

FROM: Sales Department

RE: Transactions Report:
        I.   Current Site Status
        II.   Year 2001 To Date—From Beginning to   2001
        III.   Recent Month

REPORT I: Current Site Status: From Beginning to   2001

Based on _____ Single Family Homes and _____ Town Homes

	Single Family	Town Homes	Total
Completed and Sold			
Signed Contracts			
Binders—Current			
Totals			
Total Available*			

*Includes Model Homes and Offices

REPORT II: Year-To-Date: *Only* From January 1, 2001 to   2001

	Single Family	Town Homes	Total
Closed Sales*			
Signed Contracts			
Binders—Current			
Totals			

REPORT III: Recent Month*: From March 15, 2001 to April 15, 2001

	Single Family	Town Homes	Total
Closed			
Signed Contracts			
Binders—Current			
Totals			

*Period covered is from and including date of last Monthly Report to the date of this Report.

Date of last Monthly Report: _____, 2001

USING THIS FORM

Who Prepares It and Who Updates It

The sales staff prepares the monthly transactions report and updates it monthly.

Who Uses It

The sales staff and the builder should review the monthly transactions report monthly. The construction lender may require a copy to determine the payment schedule for advances or draws.

How to Use It

The sales staff must revise the monthly transactions report carefully to ensure its accuracy. They must be mindful to adjust (subtract) for any canceled reservations or purchase agreements; add all new purchase agreements; decrease the number of reservations that converted to purchase agreements (whose number is then increased); and decrease the number of purchase agreements that converted to closings (whose number is then increased) from the preceding month's totals when preparing the latest monthly update version. It's easiest to work from the bottom up, that is, first calculating the monthly activity count, then the year-to-date figures, and, finally, the historical totals.

How to Organize It

The monthly transactions report is divided into three sections, each showing the number of closed sales, signed purchase agreements, current binders, and the total of the three categories. One section records the monthly activity, the year-to-date activity, and the historical activity of the community including currently available standing inventory as well as that which may be offered in the future. The total number of the closed sales, executed purchase agreements, current reservations, and available inventory should, of course, add up to the total number of homes offered in the community. As shown in the example, both single family homes and townhomes can be grouped on the same report.

Who Gets Copies

The sales staff, the construction department, the builder, and their construction lender, on request, should receive copies of the monthly transactions report.

Method of Transmission

The monthly transactions report may be hand delivered, faxed, mailed, or e-mailed.

Where It's Filed

The monthly transactions report should be filed in its own binder.

SCHEDULE OF FUTURE CLOSINGS REPORT

DESCRIPTION

The **Schedule of Future Closings Report** is a chronological listing of four critical dates for each sales transaction that have to be monitored regularly to ensure the timely delivery of the home.

PURPOSE

The purpose of the schedule of future closings report is to enable the sales and construction staffs to monitor the construction progress of each home so that its completion is on schedule to meet the delivery date specified in its purchase agreement. If it becomes apparent that the delivery needs to be modified, a new date should be established and the purchaser notified as soon as possible.

SCHEDULE OF FUTURE CLOSINGS REPORT

To: Construction Department

From: Sales Staff

Re: The Schedule of Future Closings Report for Your Review

Date:

This report is a chronological listing of your sales according to the closing dates in your purchase agreements. Please review your construction completion schedule and fill in your projected home delivery date. If you foresee any problems that will cause a delay in completion beyond the contract date, please advise and supply us with the new projected delivery dates.

#	Lot # or Unit #	Purchaser	Model	Contract Closing Date	Construction Delivery Date	Mortgage Expiration Date	Rate Lock Expiration Date
1.							
2.							
3.							
4.							
5.							
6.							
7.							
8.							
9.							
10.							
11.							
12.							

USING THIS FORM

Who Prepares It and Who Updates It

The sales and construction staffs should jointly prepare the schedule of future closings report, review it bi-weekly or monthly, and revise it when necessary.

Who Uses It

The sales and construction staffs use the schedule of future closings report. Approximately four or so weeks before a home's completion, the sales staff should initiate the steps called for on the preliminary closing report so that as soon as a certificate of occupancy is issued, everything is in place and everyone is prepared to close expeditiously.

How to Use It

The sales staff must regularly monitor the four critical dates on the schedule of future closings report: contract closing, construction delivery, mortgage expiration, and rate-lock expiration. If the first two do not coincide but still allow for a closing at a date acceptable to the purchaser and do not entail any financial penalty, the transaction can progress smoothly to settlement. However, if a delay in construction means that the contract delivery date will not be met, that the mortgage commitment will expire and with it a favorable rate lock, the purchaser will be upset and doubly so if they have to incur additional costs to extend their mortgage commitment and/or a favorable interest rate. The sales staff may expect that the purchaser will then ask for reimbursement of expenses or compensation at the very least. The sales staff should be prepared with a response. Additionally, the purchaser may exercise his/her right to terminate the sale altogether if a delay in construction and delivery exceeds the allowable time frame stipulated for completion in the purchase agreement.

The construction department should promptly notify the sales staff in writing of any delays in the construction schedule that will adversely impact the contract delivery date of any home. The sales staff should then immediately notify the purchaser in writing, explaining the reason(s) for the delay and the new projected delivery date.

How to Organize It

The schedule of future closings report should be organized as a chronological listing of the projected closing dates of the homes in contract with adjacent columns showing the construction department's best estimate of the delivery date, the mortgage expiration date, and the rate-lock expiration date.

Who Gets Copies

The sales staff, builder, and construction staff should receive copies of the schedule of future closings report.

Method of Transmission

The schedule of future closings report should be hand delivered or faxed.

Where It's Filed

The schedule of future closings report should be filed in its own binder.

PRELIMINARY CHECKLIST FOR CLOSINGS

DESCRIPTION

The **Preliminary Checklist for Closings** is a list of items that must be completed prior to the closing and in order for it to be scheduled.

PRELIMINARY CHECKLIST FOR CLOSINGS

Updated: _____

Lot #	Name	Survey Ordered—Received	Certifications Ordered—Received	Site Visit Letter	Certificate of Occupancy	Closing Cost Statement Mailed—Received	Pre-Closing Letter	Final Lender Inspection	Final Walk Thru	Closing Date

PURPOSE

The purpose of the preliminary checklist for closings is to remind the sales staff to take the necessary steps and secure the appropriate documentation to ensure a timely closing. Often all or at least a substantial part of this task is performed either by the builder's attorney or by a closing coordinator in the builder's office.

USING THIS FORM

Who Prepares It and Who Updates It

The sales and construction staffs prepare the preliminary checklist for closings. They determine their mutual responsibilities and coordinate their efforts to complete the essential steps preliminary to but prerequisite for closings. The sale staff updates the preliminary checklist for closings.

Who Uses It

The sales staff, the builder, and the construction staff use the preliminary checklist for closings.

How to Use It

As each step on this checklist is completed, the date is entered in the appropriate column. The closing can be held as soon as all of the items are completed. However, if one or more of these items are not attended to and a closing has been tentatively scheduled, it will not take place. To avoid the anguish, embarrassment, and costs of a delay and postponement, it's critically important that this checklist be monitored regularly, daily if required.

How To Organize It

The preliminary checklist for closings is organized as a chart itemizing a series of steps that need to be completed before the closing can take place. These steps include the issuance of a congratulatory letter to the purchaser after the purchase agreement is signed explaining the community's procedures and policies from that point on; a closing cost statement; a survey; certifications (so that title can be ordered); a certificate of occupancy; a pre-closing letter to the purchaser explaining what he/she has to do prior to closing; a final lender inspection; and a final pre-closing walk-through inspection or homeowner's orientation. Some organizations may wish to include additional items based on their company's policies and practices. Appendix F offers an even more comprehensive closing preparations checklist.

Who Gets Copies

The sales staff, the builder, and the construction staff receive copies of the preliminary checklist for closings.

Method of Transmission

The preliminary checklist for closings is transmitted by hand delivery or by fax.

Where It is Filed

The preliminary checklist for closings is filed in its own binder.

COMPARABLE SALES REPORT

DESCRIPTION

The **Comparable Sales Report** is a numerical and chronological listing of all of the closed sales with the complete financial details of each one.

PURPOSE

The comparable sales report is designed to facilitate the work of appraisers and thereby expedite the mortgage approval and commitment process.

COMPARABLE SALES REPORT

#	Unit/ Lot #	Purchaser(s)	Model	Date	Base Price +Lot Prem.	Upgrades/ Options-Credits	Total Price
1.							
2.							
3.							
4.							
5.							
6.							
7.							
8.							
9.							
10.							
11.							
12.							
13.							
14.							
15.							
16.							
17.							
18.							
19.							
20.							
21.							
22.							

USING THIS FORM

Who Prepares It and Who Updates It

The sales staff prepares the comparable sales report and updates it by recording the details after each closing.

Who Uses It

The sales staff and the appraisers use the comparable sales report. The builder and the construction staff may also find it useful as a historical and financial record of the sales. The comparable sales report should be included in the annual report discussed below.

How To Use It

The sales staff uses the comparable sales report to provide appraisers with the financial details of comparative closings to assist them in determining the value of the residence being considered for lender financing. Appraisers generally use only comparables that have closed within the previous six months. Appraisers will also ask for the subject property's floor plan with square foot measurements; a list of standard features; the options and upgrades and their pricing; a site map; the status of site amenities; and information on the size of the community, the number of closed sales, number of rented units or investor sales, and the number of unsold units. In the case of a condominium, co-operative, or HOA unit, the appraiser may request the offering plan and any amendments, information on the Board of Directors, the insurance and the common charges or community fees, and what they cover and more.

It's prudent to keep "comps" updated and handy together with the collateral materials that appraisers will request. Scheduling a closing is often contingent on the lender's receipt of the appraiser's report, which is sometimes combined with the lender's final inspection of the property.

How to Organize It

The comparable sales report provides information on each sale organized in the following categories: lot or unit number and street address; purchaser's name; model type; closing date; base price; lot premium, if any; expenditures for options and upgrades; credits; builder incentives or concessions; and final sales price.

Who Gets Copies

The sales staff and appraisers should get copies. The builder and construction staff may also request copies.

Method of Transmission

The comparable sales report should be distributed by hand or by fax.

Where It's Filed

The comparable sales report should be filed in its own binder.

ANNUAL REPORT

DESCRIPTION

The **Annual Report** is a compilation of selected forms that provides a retrospective review and summary of the past 12 months' sales activity along with a statement of the goals, projections, and recommendations for the forthcoming year.

PURPOSE

The purpose of the annual report is to provide information for analysis of the achievements and shortcomings of the past 12 months, to set goals for the future, and to determine what changes should be implemented to enhance performance and increase bottom line profitability.

USING THIS FORM

Who Prepares It and Who Updates It
The sales staff prepares the annual report for distribution shortly before or during the first week of the community's calendar or anniversary year.

Who Uses It
The annual report is used by the sales staff, the builder, and the construction staff.

How To Organize It
The annual report is a compilation of selected end-of-the-year forms that might include but not necessarily be limited to the following:

◆ The transaction (sales activity) report.

◆ A closings to date report. This is easily compiled from the comparable sales report. It also may serve as the basis for a directory of homeowners with the addition of the purchasers' telephone numbers.

◆ The schedule of future closings report.

◆ The price list(s).

◆ A report for the coming year stating goals, projections, and recommendations.

Who Gets Copies
The sales staff, the builder, and the construction staff receive copies of the annual report.

Method of Transmission
The annual report is distributed by hand or by mail.

Where It's Filed
The annual reports are filed in an annual report file.

CHAPTER 3

A PROTOCOL FOR SELECTIONS, OPTIONS, AND UPGRADES AND CHANGE ORDERS

OVERVIEW

The selection process wherein the purchasers choose the standard features and the options and upgrades that will become part of their new home can be an enjoyable, fun-filled experience or a tedious and time-consuming chore. In communities that offer limited choices and minimal options, the sales manager generally assists the buyers. In upscale or high-volume communities, a specially designated customer service representative or options coordinator, who often has a background in design, handles this function. Patience and a sense of humor are required because purchasers frequently agonize over the decisions and have second thoughts about them. Many will reconsider their choices and redo them, occasionally several times over. Spouses may strongly disagree with one another about color choices and design compatibility. Friends and relatives who are brought along to assist may only complicate and prolong the process. In the entire sales transaction, selections can truly become the "time that tries mens' souls." Whoever assists the purchasers with their selections must be well prepared for the undertaking.

DESCRIPTION AND EXPLANATION

The selection process is carried out in a designated area complete with displays, samples, and product catalogues. Occasionally, a purchaser may be given an allowance and sent offsite to a vendor's showroom to make certain selections (e.g., kitchen design, lighting fixtures, floor coverings) and will work directly with the store's representative, who will, in turn, supply the selections to the sales and construction staffs. In this chapter we provide two basic forms and formats for recording selections. In the first form, the purchaser makes the selections and the builder's representative fills in the blank spaces and calculates the prices for any nonstandard feature on the appropriate pages of a prepared form. In the second form, all of the available options are conveniently pre-priced on the prepared form and the purchaser merely has to circle the selections. In both formats the purchaser initials or signs and dates each page including a summary total page. Any subsequent change orders are consecutively numbered, signed, and dated and appended to the protocol, which is affixed to the selections section of the purchaser's file folder.

HOW THIS SYSTEM FITS WITH THE OTHER SYSTEMS

The selection process may be initiated and completed either before or preferably after the purchase agreement is executed. Once completed, there is usually a deadline stated for any future modifications by change order and an administrative fee. The purchaser's expenditures for options, upgrades, and change order items and any credits extended are itemized on the purchaser's closing cost financial schedule.

PROTOCOL FOR SELECTIONS, OPTIONS, AND UPGRADES

DESCRIPTION

Forms and formats, procedures and policies are presented in this protocol to facilitate a purchaser's selection of standard features, options and upgrades, and custom changes for the new home.

Entry-level sites offer a limited selection of standard features and options and upgrades so that choices can be completed with the sales counselor or an assistant relatively quickly—in an hour or so. High-volume and/or upscale communities, featuring a wide array of options and upgrades, commonly allow for several time-consuming sessions with a customer service representative or options coordinator (who may possess a background in design) to oversee the process.

If the sales counselor is charged with selections, with few exceptions the process should be conducted on week days (rarely on weekends, which is prime selling time) during regular business hours when construction personnel are also available for assistance if needed. The options coordinator can offer more flexible hours, including weekends, and is usually compensated in whole or part based on a percentage of the purchaser's expenditures. Experience has shown that purchasers will complete selections more quickly when their small children are not present and on week days rather than on weekends when most buyers have more leisure time.

PURPOSE

The **Protocol for Selections, Options, and Upgrades** provides a complete record of all the standard features, options, upgrades, and modifications (additions, deletions, and substitutions) selected and the charges or credits that apply.

USING THIS FORM

Who Prepares It and Who Updates It

The builder and the construction and sales staffs should jointly prepare this protocol and revise it as necessary. After the selections are completed, they may be modified by the addition of one or more change orders. The prices charged for options and upgrades should be reviewed at least annually and adjusted to reflect market conditions. Each builder has his/her own philosophy regarding selections. At one end of the spectrum are those who like to keep it simple and offer only limited choices for standard features and a small number of options and upgrades. They view the process as a necessary concession required to make sales. At the other extreme are those who offer a generous selection of options and upgrades exhibited in eye-catching displays. They see the process as an income stream and substantial profit center. Most builders take a position somewhere in the middle.

Who Uses It

The builder, the sales and construction staffs, and the purchaser use the protocol for selections, options, and upgrades.

How To Use It

The sales counselor or options coordinator reviews the selection sheets line by line, section by section, and page by page and records the choices made by the purchaser. The itemized costs on each page should be individually totaled and initialed by the purchasers. The itemized options and upgrades and their costs should be restated on the cover page and the purchasers requested to sign and date it.

We have found that most purchase agreements state that if after written notice the purchaser fails to make the selections in a timely fashion, the builder or sponsor may make the selections for him (so as not to delay construction and jeopardize the delivery date). We recommend that there be a written notice in **boldface** above the signature line on the final page to the effect that after a certain date any additional change orders are at the sole discretion of the builder, who may impose a levy to cover either the additional administrative costs or the carrying costs of the home if a delay in the projected delivery date becomes necessary.

It's been our experience that builders generally require that their purchasers pay for options and upgrades in full by separate check at the time those selections are made or before they are ordered. In those instances where a purchaser wishes to mortgage all or part of the cost of the options, the sales counselor should verify that they are qualified to do so with the mortgage officer but still require an equitable partial prepayment.

Two different methods of handling selections are illustrated. In the first and more common case, as at *Deerfield Ponds,* the sales counselor works with product samples, display boards, and different product catalogues to show purchasers the standard features and available custom options and upgrades.

*We strongly recommend that all selection sheets and change orders underscore in **bold type** that all moneys paid for options and upgrades are not refundable under any circumstances. **This caveat protects the builder in case a home loaded with expensive options and upgrades fails to close title and has to be resold well above its base price.***

The purchasers indicate their choices by filling in the appropriate blank spaces. The sales counselor notes any additional costs (for upgrades) and keeps a running subtotal on each page with a final total on the last page. Then, all of the options and upgrades are individually itemized and totaled on a cover sheet called Exhibit A. This is a time-consuming process and demands care to avoid errors in calculations (e.g., upgrade carpeting in bedroom #2, which costs $8.50 a square foot for a room 12 feet 8 inches by 14 feet 6 inches, etc. And don't forget the cut charge).

In the second example, at the *Westwind* community, the sales counselor or options coordinator works with the same sales aids, but the builder has pre-calculated and recorded the prices for all of the options and upgrades offered for every room of the home for each of the models. Purchasers merely circle their choices with their prices already printed on the selections sheets (in our example for the Northwind model) and have the sales counselor or options coordinator calculate the grand total on the final page. This will require an extensive effort on the part of the builder and the construction staff prior to the opening of the community sales program.

Builders may offer their basic models with a choice of three or four upgrade "packages"—for example, a "bronze," "silver," "gold," or "platinum" package—that successively offers purchasers the opportunity to select more extensive and expensive combinations of features that are grouped without any exchanges or substitutions permitted. The packages may contain seemingly unrelated items such as central air conditioning, a fireplace, and a whirlpool tub, for example, in the "bronze" package" or packages of bundled options may be organized around a theme or a particular room. For example, a "high-tech" package might include extra cable TV and telephone jacks, special wiring for high-speed Internet access, remote control for appliances and state-of-the-art entertainment, and security systems. A "gourmet kitchen" package might include an upgraded sink, upgraded appliances, a warming tray, and a wine rack.

How To Organize It
The protocol for selections, options, and upgrades may be organized according to any of the three methods outlined above—"fill-in-the blanks," "pre-printed prices," or "package" offerings—with or without modifications to any one of them.

Who Gets Copies
The sales staff, the construction department, and the purchaser all receive copies of the protocol for selections, options, and upgrades. The mortgagee and possibly the builder's accounting department may also request copies.

Method of Transmission
The protocol for selections, options, and upgrades is transmitted by hand delivery or fax.

Where It's Filed
The protocol for selections, options, and upgrades is filed in the purchaser's file folder in the sales office and in their file in the construction department.

PROTOCOL FOR SELECTIONS, OPTIONS AND UPGRADES
Deerfield Ponds
Exhibit A

Buyer:_____   Lot#:_____   Model:_____

Home#:_____   Work#:_____   Date:_____

Item:	Amount:
Total Amount of Upgrades	$

Purchaser's Signature:_____   Date:_____

Purchaser's Signature:_____   Date:_____

_____   Date:_____

_____   Date:_____

Page ___ of ___

Deerfield Ponds
Selections and Upgrades

Buyer:_____ Lot#:_____ Model:_____

Home#:_____ Work#:_____ Date:_____

Cabinets:

	Std or Upgrade	Color	Sytle #	Hardware	Add'l Charge
Kitchen					
Master Bath					
2nd Floor Bath					

Counter Tops:

	Style #	Color/Straight or Bull Nose	Std or Upgrade	Add'l Charge
Kitchen				
Master Bath				
2nd Floor Bath				
Powder				

Notes:_____

Appliances:

	Std or Upgrades	Color	Model #	Credit	Add'l Charge
Refrigerator					
Range					
Range Hood					
Dishwasher					
Microwave					
Washer					
Dryer					

Notes:_____

Sinks:

	Style #	Color/Straight or Bull Nose	Std or Upgrade	Add'l Charge
Kitchen				
Master Bath				
2nd Floor Bath				
Powder Bath				

Faucets:

	Style #	Color/Straight or Bull Nose	Std or Upgrade	Add'l Charge
Kitchen				
Master Bath				
2nd Floor Bath				
Powder				

Purchaser's Signature:_____ Date:_____

Purchaser's Signature:_____ Date:_____

Subtotal of page 1: $_____

Buyer:_____

Flooring:

	Std or Upgrade	Type: w/w, vinyl, ceramic, hardwood, pergo	Add'l Charge
Entry Foyer & Closet			
Powder Bath			
Master Bath			
2nd Floor Bath			
LR			
DR			
FR			
Kitchen			
Stairs			
Upper Hallway			
Master Bedroom			
BR #1			
BR #2			
BR #3			
BR #4			
Bonus Room			
Sun Room			

Padding:

Std	Add'l Charge

Carpet Color Change:

$140 per change	Add'l Charge

Notes:_____

Fireplace:

	Std or Upgrade	Mantle	Surround	Add'l Charge
FR				
LR				

Notes:_____

Railings:

	Std: Natural Oak	Upgraded Stain Color	Add'l Charge

Plumbing Upgrades:

	Add'l Charge

HVAC and Electrical:

Air Conditioning	Std or Upgrade	1 or 2 zones	Add'l Charge

Purchaser's Signature:_____ Date:_____

Purchaser's Signature:_____ Date:_____

Subtotal of page 2: $_____

Buyer:_____

HVAC & Electrical Continued

Cable and Telephone Outlets: (see attached floor plan)

	Standard	Extra outlets	Location	Add'l Charge
Cable				
Telephone				

Garage Door Opener	Add'l Charge

Recessed Lights: (See attached floor plan)

Location	Add'l Charge

Additional Electrical Items:

	Add'l Charge

Exterior:

Roof Color:

Shutters:	Color:	Std (front only)	Upgrade (entire house)	Add'l Charge

Door Color:

Trim Color: (Victorian Only)

Deck:	Add'l Charge
Size:	

Porch:	Add'l Charge
Size:	

Bilco Basement Doors:	Add'l Charge
Location:	

Glass Side Lights:	Add'l Charge

Purchaser's Signature:_____ Date:_____

Purchaser's Signature:_____ Date:_____

Subtotal of page 3: $_____

Buyer:_____

Exterior continued:

Additional Exterior Options	Add'l Charge

Architectural Fees:

	Add'l Charge
Printing Fee: $350.00	
Per Item Fee: $75.00–$250.00	

Miscellaneous Items:

	Add'l Charge

Note: Deerfield Ponds reserves the right to not allow any changes after these selection sheets have been signed. If the Sponsor agrees to allow any changes, there will be a $200.00 charge per item. <u>All monies paid for Options and Upgrades are not refundable under any circumstances.</u>

Subtotal of page 4: $_____
Credits: $_____
Total Due: $_____

Purchaser's Signature:_____ Date:_____

Purchaser's Signature:_____ Date:_____

_____ Date:_____

_____ Date:_____

PROTOCOL FOR SELECTIONS, OPTIONS, AND UPGRADES (CON'T)
Westwind
Exhibit B
Standard Selections and Options

Purchaser Name:_____   Estimated Closing Date:_____

Home Number:_____   Given to Construction On:_____

Phone Numbers:_____

     Work:_____

     Home:_____

Northwind Model

	Color	Price	Acceptance
Flooring:			
First Floor			
#1 Oak Wood Flooring		Standard	_____
(not including kitchen, bath, or laundry)			
Second Floor			
Carpeting	_____	Standard	_____
Carpeting			
Upgrade 1			
Entire Second Floor	_____	$1,000	_____
Master Bedroom	_____	$ 350	_____
Bedroom #1	_____	$ 200	_____
Bedroom #2	_____	$ 200	_____
Bedroom #3	_____	$ 200	_____
Second Floor Hallway	_____	$ 150	_____

	Color	Price	Acceptance
Upgrade II			
Entire Second Floor	_____	$1,690	_____
Master Bedroom	_____	$ 600	_____
Bedroom #1	_____	$ 600	_____
Bedroom #2	_____	$ 350	_____
Bedroom #3	_____	$ 350	_____
Second Floor Hallway	_____	$ 225	_____

	No. of Cuts	Price	Acceptance
Additional Color			
$100 Per Cut:	_____	_____	_____
Details: (attach floor plan)			

Flooring continued:

Wood Flooring—#1 Oak Strip Flooring

Entire Second Floor	$5,522	_____

Per Room

Master Bedroom	$2,025	_____
Bedroom #1	$1,120	_____
Bedroom #2	$1,120	_____
Bedroom #3	$1,120	_____
Second Floor Hallway	$ 750	_____

Stained Floors: _____ _____ _____

(stained floors are an additional charge)

(Please check pricing sheet for stained floors in individual rooms and write in price above)

	Color	Price	Acceptance

Foyer

Standard

	Color	Price	Acceptance
Tile	_____	$0	_____
(12 × 12 floor tile)			
Grout Color	_____	$0	_____
# 1 Oak Floor		$0	_____
(polyurethane only)			

Stairs:

Oak Treads with Pine Risers	Standard	$0	_____
(Oak Rails & Pine Spindels)			

Powder Room:

Floor Tile:	Color	Price	Acceptance
8 × 8 Standard Color	_____	$0	_____
Grout Color	_____	$0	_____
12 × 12 Standard Color	_____	$215	_____

Fixtures: (includes U/R pedestal sink & round toilet)

White	Standard	$0	_____
Almond	Standard	$0	_____
Heat Fan Light		$270	_____

Hall Bath (upstairs):

Floor Tile:	Color	Price	Acceptance
8 × 8 Standard Color	_____	$0	_____
Grout Color	_____	$0	_____
12 × 12 Standard Color	_____	$410	_____
Heat Fan Light		$270	_____

Hall Bath (upstairs) continued:

	Color	Price	Acceptance
Fixtures: (incl. Kohler Veracruz 60″ tub & shower. Bertch cultured marble top & Kohler Wellworth round toilet)			
White	Standard	$0	_____
Almond	Standard	$0	_____
Upgraded Color Choice	_____	$200	_____
Cultured Marble Sink	_____	$0	_____
Upgraded Sink top (cultured granite)	_____	$95	_____
Optional 2nd Sink	_____	$275	_____
Upgraded Sink Top 2nd for Sink	_____	$95	_____
Vanity Choice			
White w/Oak Pull	White Laminate Oak Recess Panel (circle one)		_____

Master Bath:

Floor Tile:	Color	Price	Acceptance
8 × 8 Standard Color	_____	$0	_____
Grout Color	_____	$0	_____
12 × 12 Standard Color	_____	$530	_____

	Color	Price	Acceptance
Fixtures: (incl. Kohler Veracruz 60″ tub & shower. Bertch cultured marble top & Kohler Wellworth elongated toilet)			
White	Standard	$0	_____
Almond	Standard	$0	_____
Upgraded Color Choice	_____	$200	_____
Cultured Marble Sink	_____	$0	_____
Upgraded Sink Top (cultured granite)	_____	$95	_____
Vanity Choice			
White w/Oak Pull	White Laminate Oak Recess Panel (circle one)		_____
Heat Fan Light		$270	_____
Optional 2nd Sink	_____	$390	_____
Upgraded Sink Top 2nd for Sink	_____	$95	_____

Master Bathroom Suite: Color   Price   Acceptance

		Color	Price	Acceptance
		_____	$6,235	_____
Includes:				
Kohler Sojourn Whirlpool (60 × 60 corner Whirlpool)			incl.	_____
	Tile Selection for Whirlpool	_____	incl.	_____
	Grout Color	_____	incl.	_____
	Second Sink		incl.	_____
	Choice of Vanity	_____	incl.	
	(write in choice)			
	Kohler Cancun Shower			
Upgraded Shower		_____	$1,200	_____
(includes neo-angle base & enclosure + 2 tile walls)				
	Select Tile for Upgraded Shower	_____	incl.	_____
	Grout Color for Shower	_____	incl.	_____

Kitchen:		Price #	Total	Acceptance
Florescent Undercabinet Fixtures		$150 ea._____	_____	_____
Fluorescent Lighting Package		$500	_____	_____
Floor		**Color**	**Price**	**Acceptance**
	Linoleum	_____	$0	_____
	12 × 12 Tile	_____	$2,200	_____
	Grout Color	_____	$0	_____
	# 1 Oak Wood Flooring	_____	$2,600	_____
	Upgraded 12 x 12 Tile	_____	$2,600	_____
Kitchen Cabinets:				
	Standard		$0	_____
	White w/Oak Pull	Solid Oak Recessed Panel		
		(circle one)		
Upgraded Cabinets			**Price**	**Acceptance**
	Fairfax	(soft maple raised panel)	$1,800	_____
	Paragon	(white thermofoil)	$2,210	_____
	Stratton Natural	(maple raised panel)	$2,210	_____
	Stratton Frost	(maple raised panel)	$2,210	_____
	Winston	(maple shaker style)	$2,600	_____
	Jamestown	(cherry raised panel)	$2,930	_____

42″ Upper Cabinet Option

(please see chart for additional charge & write in amount)

Cabinet Pulls:				
	Standard	_____	$0	_____
	Upgraded Handles	_____	$215	_____

Kitchen continued:

Cabinet Options:	Color	Price	Acceptance
Appliance Garage		$190	_____
Pullout Wastebasket		$100	_____
Hinged Tilt Trays for Sink		$45	_____
Cutlery Dividers		$30	_____
Rollout Shelves Lower Cabinets (per cabinet)		$115	_____
Kohler Single Bowl Stainless Sink	Standard	$0	_____

Upgraded Sinks:	Color	Price	Acceptance
Double Stainless Steel		$220	_____
Single Bowl Cast Iron Enamel	_____	$245	_____
Double Bowl Cast Iron Enamel	_____	$370	_____
Upgraded Faucet		$195	_____

Counters: (Formica)	Color	Price	Acceptance
Standard Countertop	_____	$0	_____
Upgraded Category	_____	_____	_____
Tiled Backsplash	_____	$345	_____
Grout Color for Backsplash	_____	$0	_____
Upgraded Corian Countertop	_____	$3,200	_____

Appliances:

(circle price if selected)	Color			Price	Acceptance
Refrigerator					
18.2 cubic foot—frost free (GE model #TBX18SAX)	W	A		$745	_____
20.6 cubic foot w/icemaker (Gemodel #TBX21CIX) or equal)	W	A		$930	_____
21.5 cubic foot Prem glass w/indoor icemaker (GE model #TBX22PCX or equal)	W/W	A/A	B/B	$1,300	_____
Dishwasher					
5 cycle, Potscrubber (GE model #GSD650X or equal)	W	B	A	$0	_____
7 cycle, 3 level Potscrubber (GE model #GSD4110 or equal)	W/W	A/A	B/B	$280	_____
Ducted Range Hood (GE model#JV345 or equal)	W	B	A	$0	_____
Microwave Oven W/W (GE model#JVM240 or equal)	A/A	B/B		$575	_____

Appliances continued:

Range

Gas

(GE model#JGBP24 or equal)	W	A		$0	_____

Gas

(GE model#JGBP35WA or equal)	W/W	A/A	B/B	375	_____

Siding:	**Color**	**Price**	**Acceptance**
Vinyl .44 Mil Siding	_____	Standard	_____

Trim Selection for Vinyl Siding: (please choose only one from below)

Vinyl	_____	Standard	_____
Wood—painted	_____	Standard	_____

Upgraded Siding:

Cedar Siding w/Cedar Trim	_____	TBD	_____

Paint:

(interior)

Two coats (1 primer, 1 white)	White	Standard	_____
Upgraded Paint			
(1 primer, 2 coast any color offered)		$2,500	_____
Wall—Flat	_____		_____
Trim—Semi Gloss	_____		_____
Ceiling	White	Standard	_____
Additional Paint Colors		$100 ea.	_____

(attach diagram from paint colors)

Additional Options Available:

Basement:	**Color**	**Price**	**Acceptance**
Roughed in plumbing only		$1,500	_____
Finished Basement with full bath		$19,000	_____
Finished Basement without full bath		$14,000	_____

(Finished basements include HVAC on a 3ʳᵈ zone, lights & outlets, finished walls & ceilings, carpet & paint)

(Only fill in the below selections if finished basements are purchased)

Carpet	_____	incl.	_____

Bathroom Fixtures:

(incl. Kohler Veracruz tub/shower, 2 foot vanity, Bertch cultured marble top & Kohler Wellworth round toilet)

White	Standard	incl.	_____
Almond	Standard	incl.	_____
Upgraded Color Choice	_____	$200	_____
Tile Selection (bath only)	_____	incl.	_____
Grout color	_____	incl.	_____

Vanity Choice

White w/Oak Pull	White Laminate	Oak Recess Panel	_____

(circle one)

Basement continued:

Fireplace: (includes mantel, hearth & slate surround)

Wood	Standard	_____
Upgraded Gs	$750	_____

Central Vacuum (1 outlet per floor) $1,500   _____

(includes 4 outlets, main unit 28′ hose, and basic attachments)

Humidifier on HVAC System $650   _____

Air Cleaner on HVAC System $650   _____

Attic Fan $700   _____

(includes louvered opening in 2nd floor ceiling & vent in gable)

Garage Door Openers $750   _____

(includes one key pad & 2 remotes)

Deck (approx. 190 sf includes rails) $3,600   _____

Deck Over Walkout Basement (approx. 190 sf includes $4,400   _____

rails & s)

Patio (approx. 140 sf) $ 980   _____

Upgraded Trim Package

(incl. 7″ decorative base molding, 3½′ door casing and solid core doors, where applicable)

Dining Room & Living Room $685   _____

Entire House $4,525   _____

First Floor $2,375   _____

Second Floor $2,200   _____

Central Intercom system $2,750   _____

(includes AM/FM master unit, 4 indoor remotes, exterior remote & 1 integrated chime)

Upgraded Lighting Package $975   _____

Alarm System $1,700   _____

Attic Stairs $1,200   _____

(includes approx. 256 sf of attic floor)

Vaulted Ceiling Master Bedroom $5,800   _____

Skylight $1,500 ea.   _____

(Master Bedroom Only)

Windows:

Double Hung Low E Vinyl Windows   Standard   _____

Anderson Double Hung Windows $3,900   _____

Anderson Tilt-Out Windows $4,800   _____

Upgraded Landscaping Package: $2,400   _____

(includes 3,000 sf of additional lawn, and plantings; please see spec sheet for details)

Cost of Options: $_____

Cost of Additional Electrical: $_____

Cost of Additional Options: $_____

Total Cost of Options: $_____

Check Amount Received: $_____

                                    Balance due at closing $_____

Agreed to and accepted this _____ day of _____

Owner/Owner's Agent: Purchaser(s):

_____   _____

Additional Electrical Options Available:

	Color	Price	#	Acceptance
Additional Telephone Jacks (per jack)	_____	$60	_____	_____
Additional Cable Jacks (per jack)	_____	$150	_____	_____
Additional Electrical Outlets (per outlet)	_____	$60	_____	_____
(attach diagram showing location of additional jacks and outlets)				
Additional Recessed Lights	_____	$120 ea	_____	_____
Recessed Light Package (4)	_____	$400	_____	_____
(off of one (1) light switch)				
Recessed Light Package (4)	_____	$460	_____	_____
(off of two (2) light switches)				
Recessed Light Package (4)	_____	$530	_____	_____
(off of three (3) light switches)				
(Please show light placement on attached diagram)				
Additional 3-way switches (2 switches)	_____	$60	_____	_____
Additional 4-way switches (3 switches)	_____	$130	_____	_____
Additional interior GFI outlet	_____	$100	_____	_____
Additional exterior GFI weather-proof outlet	_____	$120	_____	_____
Speaker outlets (per 2 speaker outlets)	_____	$175	_____	_____
Upgraded Decora services throughout house	_____	$650	_____	_____
Reinforced boxes for ceiling fan and/or				
Changliers (to replace existing box)	_____	$70	_____	_____
Additional reinforced boxes for ceiling fan				
and/or chandeliers	_____	$100	_____	_____
Total				
(please add to indicated line on option sheet)				_____

INFORMATION/PRICE REQUEST FORM

DESCRIPTION

The **Information/Price Request Form** is a memorandum from the sales staff to the construction department soliciting information.

PURPOSE

The purpose of the information/price request form is to obtain a written response from the construction department addressing concerns or inquiries posed by purchasers that commonly deal with construction specifications, policies and procedures, or the feasibility and pricing of unusual options, upgrades, or modifications to a floor plan or that relate to some aspect of the site or general community.

USING THIS FORM

Who Prepares It and Who Updates It
The sales staff prepares the information/price request form and updates it as warranted.

Who Uses It
The sales staff uses this form in conjunction with the construction department.

How to Use It
The sales staff completes the form requesting the information. Often a marked up floor plan is attached illustrating the change(s) and location(s) that are the subject of the inquiry. Once the response is received, the purchaser will decide if he/she wants to proceed. If so, the sales staff will prepare a change order form, which in turn is added to the selections compartment of the purchaser's file folder and becomes part of the purchaser's permanent record.

Typical inquiries might include:

◆ Who is responsible for snow removal? Are the walkways included?

◆ Will the roads be dedicated to the town? If so, when?

◆ Are the results of the community's well water test available?

◆ At what point will the homes have individual mail delivery?

◆ How will the grading and landscaping around the retention pond be completed?

◆ How soon after I move in should I stain my deck?

◆ Can the deck be extended two feet? If so, at what cost?

◆ Can I purchase a larger oil tank? If so, at what cost?

◆ Will you floor the entire attic? If so, at what cost?

◆ Is there a credit if a purchaser brings his/her own refrigerator? If so, what is it? Why so low?

◆ The purchaser's family is in the kitchen cabinet business. Can they install their own cabinets? If so, how do we arrange for this?

INFORMATION / PRICE REQUEST FORM

To: _____

From: _____

Date: _____

Lot#/Unit#:_____Purchaser:_____

1. _____
   _____

2. _____
   _____

3. _____
   _____

4. _____
   _____

5. _____
   _____

6. _____
   _____

7. _____
   _____

8. _____
   _____

Return Date: _____

Comments: _____

_____

_____

Signature: _____

- The purchaser requests that stereo wire be run under the carpet. Please advise.

- What is the cost of two sconce lights in the dining room?

- Will we install the purchaser's massive chandelier? If so, at what cost?

- Can the basement have a "bilco door"? If so, at what cost?

- Can a half bath be installed in the basement? If so, at what cost?

- Can you replace the basement columns with a steel I beam? If so, at what cost?

- Can the garage be insulated or heated? If so, at what cost?

- Can arrangements be made for the purchaser to install his/her own alarm system? Please advise.

- Can the washer/dryer hookups be moved from the second floor to the basement?

How To Organize It

The information/price request form should have sections to list the purchaser and their lot number or unit number, their inquiries, the responses, and other comments and, of course, the date of the request and the response and signature.

Who Gets Copies

Copies of the information/price request form should be retained by the construction department and returned to the sales staff for review with the purchaser.

Method of Transmission

The information/price request form is hand delivered or transmitted by fax.

Where It's Filed

The information/price request form is filed in the purchaser's file folder in the sales office and with the construction department.

CHANGE ORDER FORM

DESCRIPTION

The **Change Order Form** is a document transmitted from the sales staff to the construction department to confirm any deviations from the floor plan and standard features that a purchaser is making to his/her residence.

PURPOSE

The purpose of the change of order form is to record a purchaser's custom changes to his/her residence (modifications, additions, deletions, or substitutions), the cost involved, and the terms and conditions under which approved changes will be made. It is essential to have a complete record of all change of order forms to prepare an accurate closing cost financial statement.

USING THIS FORM

Who Prepares It and Who Updates It

The sales staff and the construction staff prepare the change of order form and revise it as warranted.

CHANGE ORDER FORM#_____

Change Order No. _____

Purchasers_____ Date_____

Contract Dated_____ Plan _____

Address _____ Lot no._____

Description of Change(s) Requested

_____

_____

_____

_____

_____

Administrative fee $_____   Drawing attached_____

Cost of change $_____   Delivery date adjustment_____days

Credit (deleted items) $_____   Expiration date_____

Total $_____

The change described above, its cost, and the corresponding adjustment in the construction schedule has been requested by Purchasers. By signing this change order, Purchasers agree to pay for this change and acknowledge that the estimated delivery date for the home is revised accordingly. _____ will incorporate the change into the home only if the requested change described has been approved and signed by _____ and signed and paid in full by Purchasers prior to the expiration date above. Once the expiration date has passed, _____ has the options of changing the cost and delivery date adjustment, or declaring the change requested null and void.

_____   _____

Approved   Date

_____   _____

Purchaser   Date

_____   _____

Purchaser   Date

Who Uses It

The sales and construction staffs and the purchaser use the change of order form.

How To Use It

The sales staff completes the description of the changes requested and the costs involved, the terms of their implementation, and the effect, if any, on the delivery date. This information had previously been obtained from the construction department on an information/price request form. The sales counselor or options co-ordinator and the purchasers sign and date the form, indicating their approval.

How To Organize It

Change of order forms for each purchaser should be consecutively numbered and dated and should contain lot/unit and model type information, a description of the changes and their costs, terms and conditions for implementation, the effect, if any, on the construction schedule, and provide for the signatures of all parties.

Who Gets Copies

The sales staff, the construction department, and the purchaser should receive copies of each change of order form.

Method of Transmission

Change of order forms should be conveyed by hand delivery or fax.

Where It's Filed

Each change of order form should be filed in the selections section of the purchaser's file folder in the sales office as well as in the purchaser's file in the construction department.

CHAPTER 4

FINANCE PROCEDURES, CHARTS, AND WORKSHEETS

OVERVIEW

Every sales counselor should be conversant with the variety of financial programs that are available to their prospects even if one or more mortgage officers service the site. Sales counselors should also be able to qualify their prospects and answer their questions about the downpayment, the monthly payment, and closings costs, among others, and be able to explain the tax benefits of homeownership.

Typically, several lenders will solicit the builder and the sales staff to "sit" or service the site on a regular basis and for permission to leave their materials for visitors to review. Often the construction lender will make arrangements to handle a percentage if not all of the end loans. However, with the recent proliferation of loan products and mortgage lenders, many more options are available to the consumer. This proliferation of options presents a greater challenge to homeowners and sales counselors alike. The importance of financing knowledge cannot be overstated, not only to prevent transactions from failing to close due to financial problems but also to service customers by matching the financing to their specific needs. The relatively new Residential Financial Council in St. Louis, Missouri provides special training leading to the Certified Finance Specialist designation in recognition of the critical importance of this issue.

DESCRIPTION AND EXPLANATION

Virtually all institutional lenders will provide applicants with a packet of materials that includes a residential loan application, a checklist of items that the applicant must supply, an explanation of the steps leading to final settlement, a glossary of terms, and financial worksheets. Some lenders include a "Home Affordability Guide," enabling prospects to select the downpayment percentage and then move a sliding scale chart to their yearly income figure to determine the maximum sales price that they can afford at various interest rates. This is a useful if limited device. Although many prospects obtain a prequalification letter, if not a pre-approval letter, from a lender before they start a new home search, sales counselors cannot rely on this being presented and must be prepared to provide assistance.

We present five forms that will allow a sales counselor to qualify purchasers and determine their suitability for financing. These forms are used to calculate the costs involved in making a purchase decision and to answer the most commonly asked questions, including:

◆ Do I qualify for a mortgage (given my income, assets, available funds, credit history, and employment record)?

◆ What is the maximum mortgage amount that I can qualify for?

◆ What is the maximum purchase price that I can qualify for?

◆ What would my monthly payment be (principal, interest, taxes, and insurance)?

◆ What are my closing costs?

◆ What is my tax saving (or the tax benefit of homeownership)?

HOW THE SYSTEM FITS WITH THE OTHER SYSTEMS

Before a purchaser agreement is issued, the sales counselor must be certain that the prospect is not only a ready and willing buyer but also financially able to proceed. Many communities require that purchasers authorize a credit report to be run and the results, if not an actually copy of the report, be conveyed to the sales staff. A positive outcome will ensure that the prospect is a bona fide purchaser. The credit report and the other forms presented will enable the prospect to determine and analyze the financial details involved in a purchase decision. Should the purchaser then proceed, he/she should do so fully informed and confident of his/her ability to go forward with one of the most significant investments in a lifetime. Most purchase agreements require that the purchaser make timely mortgage application, usually within 10 days of receipt of a fully executed contract.

Sales counselors must also know their builder's policies and procedures as well as the available financing programs. For example:

◆ What is the minimum acceptable downpayment at contract signing?

◆ Is a "staggered" downpayment schedule acceptable?

◆ Is the community Fannie Mae, Freddie Mac, VA, or FHA approved?

◆ Is the builder offering any financial incentives or concessions (e.g., subsidy of bank closing costs, a mortgage rate buydown)?

◆ Has the builder received special federal, state, or local grants or funding? If so, do any limitations or restrictions apply?

Finally, once a mortgage has been approved, a copy should be included in the purchaser's file folder. The mortgage expiration date and rate-lock expiration date should be noted on the schedule of future closings report.

Five forms are presented in this chapter:

◆ Credit report authorization form

◆ Analysis of (condominium) ownership cost worksheet

◆ Chart to calculate the maximum mortgage qualified for

◆ Worksheet to calculate the maximum home purchase price

◆ Chart to calculate the monthly mortgage payment

CREDIT REPORT AUTHORIZATION FORM

I authorize _____and_____to
(Lender)   (Community)

obtain a credit report on the following: (If married less than two years, or have been known by another name, please indicate. Please note if you are a Jr., Sr., or III)

BORROWER

Name:_____

Address: _____

_____Zip:_____

Social Security Number:_____

If at above address for less than 2 years, please provide prior address:

_____ Zip:_____

Signature:_____ Date:_____

CO-BORROWER   Relationship to Borrower:_____

I authorize _____and_____to
(Lender)   (Community)

obtain a credit report on the following: (If married less than two years, or have been known by another name, please indicate. Please note if you are a Jr., Sr., or III)

Name:_____

Address: _____

_____Zip:_____

Social Security Number:_____

If at above address for less than 2 years, please provide prior address:

_____ Zip:_____

ANALYSIS OF (CONDOMINIUM*) OWNERSHIP COST WORKSHEET

Community:_____	Unit No._____Bldg. No._____	
Prepared for:_____	Model _____	
Date:_____	Garage_____	
	Outside space_____	
	Price $_____	
Downpayment	$	
Mortgage _____yrs. @_____%	$	
Monthly Interest and Amortization	Deductible $	$
Monthly Real Estate Taxes (Est.)	$	$
Monthly Common Charges (Est.)	XXXX	$
Total Monthly Payments (Est.)	XXXX	$
**Monthly Charges Deductible for Income Tax Purposes %		-$
Net Monthly Cost after Income Tax Deduction (Est.)		$

Est. Total Mo. Payment $_____ × 12 = $ _____ ÷ .28 = $_____
Required annual income

Est. Closing Costs:_____. Closing costs vary significantly depending on the type of residence, on the type and amount of the loan, on the lender's fees, on the date of the closing, etc.

---

*The Worksheet may be easily modified for a Single Family, Co-op, or HOA. purchase. It uses the traditional 28%–36% ratios of housing cost and debt (8%) to gross income. The lender may show considerable leeway depending on the type of loan, on the amount of downpayment, on the borrower's occupation, on whether it's a portfolio or warehoused loan or to be immediately packaged and sold in the secondary market, etc.

**The Sales Counselor should be familiar with federal and state (if any) income tax brackets.

CHART TO CALCULATE THE MAXIMUM QUALIFIED MORTGAGE

This chart is offered as a guide to help you determine how large a mortgage you might qualify for based on your annual income. The chart uses a conservative 25% ratio (instead of the standard 28%) because it assumes that the amount you need to set aside to pay for taxes and insurance would amount to approximately the 3% difference. However, many first-time homebuyer programs use higher qualifying ratios and require less income than what is listed here.

Annual Income

Interest Rates	$15,000	$20,000	$25,000	$30,000	$35,000	$40,000	$45,000	$50,000	$55,000	$60,000	$65,000	$70,000
6.5%	49,400	65,900	82,400	98,800	115,300	131,800	148,300	164,800	181,300	197,700	214,200	230,700
7.0%	47,000	62,600	78,300	93,900	109,600	125,300	140,900	156,600	172,300	187,900	203,600	219,200
7.5%	44,600	59,600	74,500	89,400	104,300	119,200	134,100	149,000	163,900	178,800	193,700	208,600
8.0%	45,000	56,700	70,900	85,100	99,300	113,500	127,700	141,900	156,100	170,300	184,500	198,700
8.5%	40,600	54,100	67,700	81,200	94,800	108,300	121,900	135,400	149,000	162,500	176,100	189,600
9.0%	38,800	51,700	64,700	77,700	90,600	103,500	116,500	129,400	142,400	155,300	168,200	181,200
9.5%	37,200	49,500	61,900	74,300	86,700	99,100	111,400	123,800	136,200	148,600	161,000	173,400
10.0%	35,600	47,400	59,300	71,200	83,000	94,900	106,800	118,600	130,500	142,400	154,300	166,100
10.5%	34,200	45,500	56,900	68,300	79,700	91,100	102,400	113,800	125,200	136,600	148,000	159,400

WORKSHEET TO CALCULATE THE MAXIMUM HOME PURCHASE PRICE

Payment Factor Table

Interest Rate	15-Year Factor	30-Year Factor
6.000	8.44	6.00
6.250	8.57	6.16
6.500	8.71	6.32
6.750	8.85	6.49
7.000	8.99	6.65
7.250	9.13	6.82
7.500	9.27	6.99
7.750	9.41	7.16
8.000	9.56	7.34
8.250	9.70	7.51
8.500	9.85	7.69
8.750	9.99	7.87
9.000	10.14	8.05
9.250	10.29	8.23
9.500	10.44	8.41
9.750	10.59	8.59
10.000	10.75	8.78
10.250	10.90	8.96
10.500	11.05	9.15
10.750	11.21	9.33
11.000	11.37	9.52
11.250	11.52	9.71
11.500	11.68	9.90
11.750	11.84	10.09
12.000	12.00	10.29
12.250	12.16	10.48
12.500	12.33	10.67
12.750	12.49	10.87
13.000	12.65	11.06
13.250	12.82	11.26

Gross income (before taxes) each year $\rule{2cm}{0.4pt}$ (A)
(for self and any co-borrowers)

Enter your gross monthly income (A ÷ 12) = $\rule{2cm}{0.4pt}$ (B)
Calculate maximum home payment[1] (B × .28) = $\rule{2cm}{0.4pt}$ (C)

Calculate the maximum debt payment[1] (B × .36) = $\rule{2cm}{0.4pt}$ (D)

Enter total monthly debt payment $\rule{2cm}{0.4pt}$ (E)
(credit cards, car payment, child support, alimony, etc.)

Calculate maximum available monthly house $\rule{2cm}{0.4pt}$ (F)
payment[2] (D−E) =

Your maximum available monthly house payment $\rule{2cm}{0.4pt}$ (G)
(the estimate you should use is the lower of C or F)

Calculate your estimated maximum mortgage loan amount:

$\rule{2cm}{0.4pt}$ ÷ $\rule{2cm}{0.4pt}$ % 1,000 = $\rule{2cm}{0.4pt}$ (H)
(enter G) (payment factor[3])

Enter total cash available for home purchase $\rule{2cm}{0.4pt}$ (I)

Calculate closing costs[4] (H % × .05) = $\rule{2cm}{0.4pt}$ (J)

Calculate cash available for down payment[5] (I−J) = $\rule{2cm}{0.4pt}$ (K)

Calculate your maximum home purchase $\rule{2cm}{0.4pt}$ (L)
price (H + K) =

[1]These percentages are used as industry standards. Percentages may vary.
[2]Monthly house payment includes principal, interest, taxes, and insurance.
[3]Use the payment factor table above to find the interest rate you expect to pay and choose a loan term from the top of the table. Enter the factor (where the loan term and interest rate intersect) above.
[4]Closing costs can vary, we suggest using 5% to 8% for estimates in major metropolitan areas; other areas can estimate 3%.
[5]If this amount is less than 20% (or in some cases less than 15%) of your estimated maximum mortgage loan amount, private mortgage insurance will be required. See your loan officer for details.

CHART TO CALCULATE THE MONTHLY MORTGAGE PAYMENT
(Principal and Interest)

The monthly mortgage payment is determined by the loan amount, by the term of the loan, and by the interest rate. The chart shows you the monthly payment for each thousand dollars borrowed at different interest rates.

For example, if the interest rate on a 30-year fixed rate mortgage is 7%, the monthly payment would be $6.65 for every $1,000 borrowed. If you are interested in obtaining a $100,000 mortgage, your monthly payment at 7% would be $665.00. Monthly payments for insurance and real estate taxes will be added to this figure. (e.g., 100 × $6.65 = $665.00)

Interest Rate	Monthly Payment Per Thousand Dollars			Interest Rate	Monthly Payment Per Thousand Dollars		
	15 Years	20 Years	30 Years		15 Years	20 Years	30 Years
4.25%	$7.53	$6.21	$4.93	9.25%	$10.29	$9.16	$8.23
4.5%	7.65	6.34	5.08	9.5%	10.44	9.32	8.41
4.75%	7.78	6.48	5.22	9.75%	10.60	9.49	8.60
5.0%	7.91	6.60	5.37	10.0%	10.75	9.66	8.78
5.25%	8.04	6.74	5.53	10.25%	10.91	9.82	8.97
5.5%	8.18	6.89	5.69	10.5%	11.06	9.99	9.15
5.75%	8.31	7.03	5.84	10.75%	11.22	10.16	9.34
6.0%	8.44	7.17	6.00	11.0%	11.37	10.33	9.53
6.25%	8.58	7.32	6.16	11.25%	11.53	10.50	9.72
6.5%	8.72	7.46	6.33	11.5%	11.69	10.67	9.91
6.75%	8.85	7.61	6.49	11.75%	11.85	10.84	10.10
7.0%	8.99	7.75	6.65	12.0%	12.01	11.02	10.29
7.25%	9.13	7.90	6.82	12.25%	12.17	11.19	10.48
7.5%	9.27	8.06	6.99	12.5%	12.33	11.37	10.68
7.75%	9.41	8.21	7.16	12.75%	12.49	11.54	10.87
8.0%	9.56	8.36	7.34	13.0%	12.66	11.72	11.07
8.25%	9.70	8.52	7.51	13.25%	12.82	11.90	11.26
8.5%	9.85	8.68	7.69	13.5%	12.99	12.08	11.46
8.75%	9.99	8.84	7.87	13.75%	13.16	12.26	11.66
9.0%	10.14	9.00	8.05	14.0%	13.32	12.44	11.85

DESCRIPTION

The five forms presented above will allow the sales counselor to qualify purchasers and determine whether they are suitable candidates for financing, their closing costs, their monthly outlays, and the tax benefits of ownership. Specific charts or worksheets are provided to calculate the maximum mortgage and maximum purchase price a prospect can qualify for as well as his/her monthly mortgage payment.

PURPOSE

The purpose of these forms is to determine the financial ability of a prospect to obtain financing and to supply him or her with a breakdown of the total costs.

USING THESE FORMS

Who Prepares Them and Who Updates Them
These forms are commonly found in various real estate publications and books on finance. Of course, the sales counselor will not have to use each form with each prospect, but it's useful to have them all available.

Who Uses Them
The sales counselor uses these forms to qualify prospects and assist them in making a purchase decision.

How To Use Them
Each of these forms includes its own easy, straightforward directions.

How To Organize Them
Experienced sales counselors often prefer to keep the financial forms either in a separate binder or in each file folder of the unsold lots or units together with a blank reservation form and other useful materials.

Who Gets Copies
The sales counselor should provide each prospect with a copy of the forms used to assist him/her and access his/her situation. The lending institution will require a copy of the credit report authorization form.

Method of Transmission
The finance forms are given to the prospect. The sales counselor should attach copies to the prospect information card if the prospect is "thinking it over" and may return at a later date or attach copies to the prospect's file folder if he/she proceeds to complete a reservation.

Where They Are Filed
Blank finance forms are filed in their own binder and often in the file folders of the unsold inventory. Completed forms are filed in the purchaser's file folder.

Chapter 5

ANCILLARY AND COLLATERAL ITEMS

OVERVIEW

Throughout the sales process there is routine correspondence that sets forth company policies and procedures and anticipates the questions and concerns that commonly arise. There are also less common but predictable occurrences that require forms for special use. Based on our experience, we have provided several forms to meet these needs. Of course, there may also be wide-ranging correspondence dealing with customer service issues both before and after the closing. More than eighty communiques are presented in Carol Smith's admirable and comprehensive *Dear Homeowner: A Book of Customer Service Letters* that address those issues.

DESCRIPTION AND EXPLANATION

This chapter presents a cover letter sent in response to a brochure and information request; two letters to the purchaser's attorney (that accompany the purchase agreement when it's initially sent out for review and signature and when it's returned fully executed); two letters to the purchaser (post-contract and pre-closing); riders* for sales contingencies; an early occupancy agreement; and a protocol for a "courtesy hold" (when a purchaser is ready and willing to complete a reservation but forgot to bring a checkbook). The first five forms are routinely used while the last three are less frequently but not uncommonly used.

HOW THIS SYSTEM FITS WITH THE OTHER SYSTEMS

The forms presented in this chapter will facilitate various steps in the sales process. The letters addressed to the purchaser will, respectively, outline the community's policies and procedures after the purchase agreement is signed and specify the steps that need to be taken to ensure a timely and seamless closing.

COVER LETTER FOR INFORMATION REQUEST

DESCRIPTION

The **Cover Letter for Information Request** accompanies all initial requests for brochures and other community and area information.

---

*All legal documents, including any riders or addenda to a purchase agreement or a pre-occupancy agreement, should be prepared and reviewed by the builder's attorney.

COVER LETTER FOR INFORMATION REQUEST

Enclosed please find the information you requested on _____

(community)

_____, our newest* residential community.

(location)

Ours is truly a unique community offering _____

_____

_____.

We would be happy to provide you with further information or schedule an appointment at your convenience to preview and discuss our community in further detail. Our office hours are: _____ We thank you for your interest in _____ and look forward to hearing from you shortly.

      Sincerely yours,

*If no longer the "newest," one might substitute "premier," "finest," "celebrated," "best-selling," "fast-selling," "highly desirable," "affordable," "acclaimed," etc.

Note: A Prospect Information Card should be completed for everyone who calls in for community information and then entered into the alphabetical file card system. That way, if the prospect shows up at a later date with a realtor, or a realtor tries to register that prospect by phone, you have evidence of direct, prior contact with the office. You will then have to decide whether to "honor" the realtor or perhaps suggest a lesser referral fee than the usual co-brokerage fee.

PURPOSE

The purpose of the cover letter for information request is to offer interested prospects some appealing information about the community and suggest that the prospect schedule a convenient appointment.

USING THIS FORM

Who Prepares It and Who Updates It
The sales staff prepares the cover letter for information request and updates it as warranted.

Who Uses It
The sales staff uses the cover letter for information request when responding to requests for community information that are not made onsite in person, such as telephone and e-mail requests.

How to Use It
The sales staff should immediately complete a prospect information card for anyone who is sent information about the community with the accompanying cover letter for information request. The sales counselor should also inquire how the prospect learned about the community to ascertain how to fill out the weekly traffic and sales report (recording the numbers and sources of inquiries on a daily basis) and find out if there was any realtor involvement. If the prospect says they found your advertisement in a local magazine and a week later shows up with a realtor, you will at least have evidence of prior contact and will have to decide whether to recognize the realtor and, if so, on what basis.

How To Organize It
The cover letter for information request should be a short letter with the enclosures requested.

Who Gets Copies
The sales counselor should retain a copy of the cover letter information request and the prospect information card for future follow-up if the prospect does not schedule an appointment or visit the site in a timely fashion.

COVER LETTER FOR THE PURCHASE AGREEMENT

DESCRIPTION

The **Cover Letter for the Purchase Agreement** accompanies all purchase agreements.

PURPOSE

The purpose of the cover letter for the purchase agreement is to provide the purchaser's attorney with procedures to follow on receipt and review of the purchase agreement. Of special note are the directions for making proposed changes to the document.

USING THIS FORM

Who Prepares It and Who Updates It
The builder's attorney in conjunction with the sales staff prepares the cover letter for the purchase agreement and updates it as warranted.

COVER LETTER FOR THE PURCHASE AGREEMENT

Date

Attorney's Name
Attorney's Address

Re:_____ To_____ Lot #/ Unit_____
Model Type_____

Dear Mr.:_____

Enclosed please find the Purchase Agreement in the above referenced transaction in quadruplicate. Kindly review the Agreement with your client(s) and have them sign the four copies as well as the Warrantee and Exhibit A (Selections, if completed and included at this time) Sections. We would appreciate the return of the documents within ten business days of your receipt.

Please return all four Purchase Agreements to me at:_____ along with your client's downpayment check in the amount of $_____ made payable to the order of "_____, as attorney." Once the Agreements are executed, I will return two counterparts to you.

If you have any questions or desire to make any changes to our Agreement, please call me at: _____ to discuss your proposed changes. Under no circumstances will our attorney or builder accept lengthy or extensive strikeouts to our Agreement or an extensive Rider to our Agreement. Should you have any comments, please include them on a separate Rider and not on the body of the Agreement.

Sincerely,

_____

(Name)

_____

(Title)

Who Uses It

The purchaser's attorney uses the cover letter for the purchase agreement for guidance in the review and execution of the purchase agreement with the client.

How To Use It

Once the cover letter for the purchase agreement and the purchase agreement are transmitted to the purchaser's attorney, the sales staff should follow up to see if the documents were received and if an appointment with the purchaser has been scheduled for review and signing.

How To Organize It

The cover letter for the purchase agreement should be a short letter with the enclosed purchase agreement. It should state what is included, what is requested (i.e., what steps need to be taken), a procedure for making changes, if any, and a time frame for completion and return.

Who Gets Copies

Only the purchaser's attorney will get a copy.

Method of Transmission

The cover letter for the purchase agreement is mailed to the purchaser's attorney along with the purchase agreement. Overnight mail is preferred or hand delivery if the attorney is local.

Where It's Filed

The cover letter for the purchase agreement should be filed either in the correspondence or in the purchase agreement section of the purchaser's file folder.

COVER LETTER FOR THE RETURN OF THE PURCHASE AGREEMENT

DESCRIPTION

The **Cover Letter for the Return of the Purchase Agreement** accompanies the return of all purchase agreements.

PURPOSE

The purpose of the cover letter for the return of the purchase agreement is to acknowledge the receipt and the return of the executed purchase agreement and to provide the purchaser's attorney with information regarding homeowners insurance, title, and survey to facilitate the closing.

USING THIS FORM

Who Prepares It and Who Updates It

The builder's attorney in conjunction with the sales staff prepares the cover letter for the return of the purchase agreement and updates it as warranted.

Who Uses It

The purchaser's attorney uses the cover letter for the return of the purchase agreement for guidance in preparation for the closing.

COVER LETTER FOR THE RETURN OF THE PURCHASE AGREEMENT

Date

Attorney's Name
Attorney's Address

Re:_____ To_____ Lot #/ Unit_____
      (Community)   (Purchaser)
Model Type_____

Dear Mr.:_____

Enclosed please find two fully executed copies of the Purchase Agreement in the above referenced transaction. Please forward one copy to your client.

As the closing date nears, the following information will help to expedite the process. The title company below is offering a title search at a discount price for our purchasers. We recommend that you use it.

The Title Company:

Name
Address
Phone Number
Contact Person:

The Homeowners Insurance Company:

Name
Address
Phone Number
Contact Person:

The Survey Company:

Name
Address
Phone Number

There is a set fee for the survey, and our construction department will order it. We will contact the homeowners insurance company to obtain an insurance policy naming your client(s) and the lending institution (if any).

If you have any questions, please feel free to contact me at
Sincerely,

_____

(Name)

_____

(Title)

ANCILLARY AND COLLATERAL ITEMS

How To Use It

The cover letter for the return of the purchase agreement is used to convey useful information to the purchaser's attorney with the return of the purchase agreement.

How To Organize It

The cover letter for the return of the purchase agreement should be a short letter containing information that the purchaser's attorney will need to have to obtain specific documents required for the client's closing, including the title, survey, and insurance.

Who Gets Copies

Only the purchaser's attorney will receive a copy.

Method of Transmission

The cover letter for the return of the purchase agreement is mailed to the purchaser's attorney with the return of the purchase agreement. Two-day overnight mail is preferred or hand delivery if the attorney is local.

Where It's Filed

The cover letter for the return of the purchase agreement should be filed either in the correspondence or in the purchase agreement section of the purchaser's file folder.

CONGRATULATORY LETTER TO THE PURCHASER

DESCRIPTION

The **Congratulatory Letter to the Purchaser** is forwarded as soon as the purchase agreement is fully executed. It serves multiple purposes.

PURPOSE

The purpose of the congratulatory letter to the purchaser is to explain the community's procedures from the time contracts are signed until the closing. It covers the selections process, site visitations, inspections, inquiries, and notice of future correspondence including a pre-closing letter and closing cost financial schedule. The letter is designed to put the purchaser at ease and to prepare him/her for what follows.

USING THIS FORM

Who Prepares It and Who Updates It

The builder and the sales and construction staffs should jointly prepare the congratulatory letter to the purchaser and updated it as warranted.

Who Uses It

The builder and the sales and construction staffs use this letter to inform purchasers about the community's procedures and policies to ensure the safety and well-being of all involved and to open a channel for regular communications as construction commences and progresses toward the closing.

How To Use It

The congratulatory letter to the purchaser is used as an educational device and as a reference in case a purchaser fails to observe the community's policies and procedures that the letter outlines.

CONGRATULATORY LETTER TO THE PURCHASER

Date

Mr. and Mrs._____

Address

Dear Mr. and Mrs._____

Congratulations on the purchase of your new home at _____.

We share your pride and excitement and look forward to welcoming you and your family to our community of new homeowners. For your comfort and convenience, we would like to familiarize you with our procedures and protocols as we move forward together toward your move-in date. Our goal is to ensure your peace of mind and safety as we strive to provide a seamless transition from now to your closing and to maximize your pleasure and minimize the anxiety that sometimes accompanies the process.

Your attorney will be forwarding a copy of the Purchase Agreement and our mortgage officer will assist you in completing your application. Shortly, you will be receiving a call to schedule an appointment with our customer service representative for your selections. This must be done on week days during our regular hours. We recommend that you leave small children at home. You may take all of the time (within reasonable limits) needed to complete your selections, options, and upgrades; however, construction will not begin on your new home until the selections are 100% completed (even though a Purchase Agreement has been signed).

Once your selections are completed and construction has commenced, you should compile a list of any new questions and requests, if any, for changes or additional options or upgrades and call our customer service representative directly. While your home is being built, you will be called several times by members of our construction staff for a walk through. At these specific visits ONLY—at completion of framing, at completion of sheetrock and taping, and at the final pre-closing inspection(s) —you may present a "Punchlist" of items to the staff. Of course, you may make additional visits to the home, but we ask that you schedule these visits in advance with our construction department since our Purchase Agreement provides that you must be accompanied by a representative of the seller. Please show up for these appointments on time because several are scheduled back to back. You may bring a licensed engineer or certified home inspector to preview the construction program at the first or second scheduled visit when the construction detail is open to viewing, although you must verify this with us so that appropriate time will be allotted. At the final homeowner's orientation however, ONLY THE PURCHASERS may inspect the home and provide a final "Punchlist" or sign an affidavit of acceptable completion. At that time, all of the systems will be explained and you should be assured that your new home is complete and ready for you and your family.

Unscheduled visits into the home are inappropriate and sometimes unwise. There are times when site blasting is scheduled and other times when access is undesirable, for example, when there are live electrical wires, or tile is being grouted, or polyurethane floor coatings are being applied or painting is under way or numerous contractors are busily working on the home. Additionally, OSHA (Occupational Safety and Health Administration) regulations, which govern construction sites as well as the mandate of our insurance carrier, prescribe that no one is allowed to walk through any home under construction except by appointment and accompaniment with the builder's authorized representative. Please understand, therefore, that if this is observed in the breach, and you are requested to leave, that directive comes from the owner and you will be in default of contract. Finally, neither purchasers nor their agents are allowed to do any work in the home prior to closing.

During the course of construction, local and state officials will be monitoring the building process and are required to undertake and approve a series of inspections in order for construction to progress from one step to the next. These inspections include:

1. A footings inspection
2. A framing inspection
3. A rough plumbing inspection
4. An insulation inspection
5. A rough electrical inspection
6. A final electrical inspection
7. A final inspection by the local building department prior to the issuance of a Certificate of Occupancy required for closing.

Your home is being built in conformity with local, state, even national building codes to ensure your comfort and safety. Please do not stop by the construction office with concerns and questions about the progress or nature of the construction of your home. All inquiries should be in writing and conveyed directly to the sales office.

About ten days before the closing, your home will be "keyed-off," that is, locked, so that no unauthorized person will have access to the home and the installations will be safe and secure and their new condition respected. This means that neither you nor the sales staff will have access to the premises even on the weekends (whether or not construction personnel are present).You will also receive a pre-closing letter with instructions and suggestions and a financial accounting of your purchase.

Thank you in advance for your understanding and co-operation. We look forward to your officially becoming a new neighbor at our community and we wish you many years of happiness in your new home!

How To Organize It
The congratulatory letter to the purchaser should be organized into as many sections and subjects as the builder and the sales and construction staffs deem appropriate. Subjects might include a congratulatory message, selections, visits to the site, inspections, the building process, and notice of future correspondence, particularly about closing procedures.

Who Gets Copies
The purchaser receives a copy of the congratulatory letter to the purchaser.

Method of Transmission
The congratulatory letter to the purchaser is mailed to the purchaser.

Where It's Filed
A copy of the congratulatory letter to the purchaser is filed in the section of the purchaser's file folder.

CONTINGENT SALES AGREEMENT

DESCRIPTION

The **Contingent Sales Agreement** is a rider or addendum to the purchase agreement signed by both the purchaser and the seller.

PURPOSE

The purpose of the contingent sales agreement is to provide the terms and conditions whereby a builder (the seller) recognizes that the sale to a purchaser is contingent on that purchaser's ability to sell his/her current residence and the purchaser also agrees to the terms and conditions set forth. Various alternatives are stipulated in the event that the builder finds another buyer before the purchaser is able to sell his/her current residence.

USING THIS FORM

Who Prepares It and Who Updates It
The builder's attorney should prepare the contingent sales agreement and update it as warranted.

Who Uses It
The sales staff uses this form when and if the builder permits its use in the community. This is a business decision that might be entertained on a case-by-case basis depending on a review of the circumstances.

How To Use It
The builder and their sales staff should review the request for a contingent sales agreement very carefully, weighing the pros and cons of removing particular inventory from the market for even a short period of time while the contingency is in effect.

How To Organize It
The builder's attorney should organize the form and include those stipulations that the client will accept.

Who Gets Copies
The prospective purchaser, the attorney, the builder, and the sales staff should receive copies of the contingent sales agreement.

CONTINGENT SALE AGREEMENT
(AS A RIDER TO THE PURCHASE AGREEMENT)

Since the Purchasers have indicated that this transaction is contingent upon their ability to sell their current home at_____

_____

_____

      (Address)

and the Seller has agreed to this contingency, the Purchaser and Seller also agree that should the Purchasers not have a bona fide contract of sale on their home by _____ , and

                                                  (Date)

should the Seller find another purchaser for said home _____

                                      (Address / Lot #/ Model)

the Seller will then notify the Purchasers that they have 48 hours to waive the home sale contingency, or else they will lose the opportunity to purchase this home, which may then be sold to another.

In the event the Purchasers do not waive the home sale contingency, they must notify the Seller in writing (certified mail, return receipt requested) to obtain the return of the contractual deposit money in full.

The third alternative is that if the Purchasers do not waive the home sale contingency, they may select a new home location (same model) mutually satisfactory to both Purchasers and Seller, to be completed at a later date, or an extension of the terms may be negotiated.

_____   _____

(Purchaser)   (Seller)

_____

(Purchaser)

Method of Transmission

The contingent sales agreement should be appended to the purchase agreement and returned with it by mail or hand delivery if the attorney is local.

Where It's Filed

As part of the purchase agreement, the contingent sales agreement is filed in the purchase agreement section of the purchaser's file folder.

PURCHASER'S MORTGAGE CONTINGENCY RIDER TO THE PURCHASE AGREEMENT

DESCRIPTION

The Purchaser's Mortgage Contingency Rider to the Purchase Agreement is an addendum to the purchase agreement signed by both the purchaser and the seller.

PURPOSE

The purpose of the purchaser's mortgage contingency rider to the purchase agreement is to provide the terms and conditions whereby a builder (the seller) recognizes that the sale to a purchaser is contingent on the purchaser obtaining mortgage approval within a specified time period. Various alternatives are stipulated if the contingency is not met.

USING THIS FORM

Who Prepares It and Who Updates It

The builder's attorney should prepare the purchaser's mortgage contingency rider to the purchase agreement and update it as warranted.

Who Uses It

The sales staff uses this form when and if the builder permits its use in the community. This is a business decision that might be entertained on a case-by-case basis depending on a review of the circumstances.

How to Use It

The builder and the sales staff should review the request for a purchaser's mortgage contingency rider to the purchase agreement very carefully, weighing the pros and cons of removing particular inventory from the market for even a short period of time while the contingency is in effect.

How To Organize It

The builder's attorney should organize this form and include those stipulations that the client will accept.

Who Gets Copies

The prospective purchaser, the attorney, the builder, and their sales staff should receive.

Method of Transmission

The purchaser's mortgage contingency rider to the purchase agreement should be appended to the purchase agreement and returned with it by mail or hand delivery if the attorney is local.

PURCHASER(S)' MORTGAGE CONTINGENCY RIDER TO PURCHASE AGREEMENT

Seller: _____

and

Purchasers: _____

_____

Lot No:_____Model No:_____ Bldg. No:_____

Upon the execution of this Purchase Agreement by Seller, purchaser(s) shall have seven days in which to sell the residence located at:

_____

This contingency shall be satisfied, and this Purchase Agreement no longer will be contingent upon the mortgage approval of purchaser's purchaser, on or before the earlier of _____ as may be extended pursuant to other provisions of this rider, or the purchaser's purchaser no longer has a mortgage contingency for the purchase of the above-listed house. It being understood that purchaser currently is in contract to sell the above-listed home. Purchaser(s) shall disclose all purchase offers and any other information pertaining to purchaser(s) sales efforts which Seller requests. If a sales contract is not executed within the seven (7) day period, Seller may:

(a) cancel this Purchase Agreement and return the Deposit made hereunder to purchaser(s) or

(b) grant one or more seven-day extensions of this purchaser's Mortgage Contingency Rider. Seller may impose additional requirements as a condition of any extension of time.

Purchaser(s)' failure to diligently market and show the residence or to provide the Seller with information requested within the time periods specified above shall be construed by all parties to be a waiver of this and any mortgage contingency; all other terms of the Purchase Agreement shall remain in full force and effect.

This purchaser(s) Mortgage Contingency Rider will not be effective until accepted and signed by Seller.

Purchaser(s):_____Date:_____

Purchaser(s):_____Date:_____

Seller:_____

By:_____

Purchaser's Mortgage Contingency Rider valid only until _____

Where It's Filed

As part of the purchase agreement, the purchaser's mortgage contingency rider to the purchase agreement is filed in the purchase agreement section of the purchaser's file folder.

EARLY OCCUPANCY AGREEMENT

DESCRIPTION

The **Early Occupancy Agreement** (sometimes referred to as a "pre-occ") is a legal document between the seller and the purchaser whereby the purchaser is permitted to occupy the premises or at least move some of his/her possessions into the home or a specified portion thereof prior to closing under stipulated terms and conditions.

PURPOSE

The early occupancy agreement is a seller's accommodation to a purchaser to allow him/her to move into the premises prior to the closing. In our experience it is rarely used, and we have found that most builder's attorneys will advise against it.

USING THIS FORM

Who Prepares It and Who Updates It

The builder's attorney should prepare the early occupancy agreement and update it as warranted.

Who Uses It

The builder and their sales staff use this form when and if the builder permits its use in the community. This is a business decision that might be entertained on a case-by-case basis depending on a review of the circumstances. A builder might decide to enter into an early occupancy agreement if the closing was delayed because the home was not ready and the purchaser had already sold his/her old home and has no alternative other than to incur the expense of placing personal possessions in storage and moving his/her family into temporary rental quarters.

How To Use It

The builder and their sales staff should review the request for an early occupancy agreement with their attorney very carefully. Sometimes the request allows the purchaser to move his/her possessions into a portion of the new home (e.g., the garage, basement, or lower level) while not permitting the purchaser to sleep there.

How To Organize It

The builder's attorney should organize this form. It commonly includes those stipulations that, of course, protect the client, including provisions for a deposit; adjustments for water, sewer, utilities, and taxes; an indemnification and hold harmless paragraph; a requirement for the purchaser to obtain liability insurance; stipulations detailing what happens if the premises fail to close by the agreed upon date; and more.

Who Gets Copies

The prospective purchaser, his/her attorney, the builder, and their sales staff should receive copies of the early occupancy agreement.

Method of Transmission

The early occupancy agreement can be mailed, faxed, or hand delivered.

EARLY OCCUPANCY AGREEMENT

THIS AGREEMENT, made the_____day of 200_, BETWEEN_____

and_____ (the "Purchasers") and_____ (the "Seller").

Purchaser and Seller have entered into that certain Purchase Agreement dated_____, for the premises

at _____("Premises").

Purchaser has requested that the Seller allow purchaser to move into the Premises prior to the occurrence of the actual date of closing.

Subject to the terms and conditions hereof, Seller has agreed to purchaser's request to allow such early occupancy.

NOW, THEREFORE, in consideration of the premises and for other good and valuable consideration, the receipt and sufficiency of which is hereby acknowledged, the parties hereby agree as follows:

1. Seller has agreed to allow Purchaser to occupy the Premises commencing on _____. In the event that the closing does not take place on _____ due to any act or omission of the Purchaser, in addition to any other right of the Seller under the Purchase Agreement, the Seller shall be entitled to, and the purchaser hereby agrees to pay, an occupancy fee of ($) for a period commencing _____ until closing, payable each day in advance based on the actual number of days that the purchaser occupies the premises prior to the closing. Notwithstanding the forgoing, in the event that the Purchaser does not close on or before thirty (30) days from _____, Purchaser without notice agrees to remove their belongings from the Premises and to vacate the Premises. Purchaser's agreement to so vacate is a material inducement to Seller to allow purchaser early occupancy.

2. Purchaser hereby unconditionally and irrevocably agrees to indemnify and hold Seller harmless from and against any and all claims of liability and/or damage to the Premises which may arise as a result of purchaser, or purchaser's agents, licensees or invitees, entering and or using the Premises prior to closing. The aforesaid indemnity shall include any injury to any person or property occurring on or about the Premises or as a result of any activity being conducted on the Premises, whether or not such activity has been specifically authorized or directed by the Purchaser. In addition the aforesaid indemnity shall include any and all reasonable cost or attorneys' fees incurred by the Seller in maintaining or defending any action in law or equity with respect thereto.

3. Prior to assuming possession, purchaser shall deliver to Seller a certificate evidencing liability insurance in an amount of $_____ naming Seller as additional insured.

4. Prior to delivery of possession, purchaser has deposited the sum of $_____ with Seller's attorney to be held in escrow pursuant to the Purchase Agreement dated _____. Notwithstanding anything contained in the Purchase Agreement to the contrary, the escrow agent may apply this sum to (a) the repair of any damages to the Premises made by the purchaser incurred after the date of delivery of possession; (b) to the payment of any taxes, water charges, or utility charges incurred after said date; or (c) to the payment of any other amounts which are the responsibility of the Purchaser under the terms and conditions of this Agreement, following the failure of the Purchaser to pay for any of the above items upon demand. In the event the escrow is utilized in the manner provided in this paragraph, then the amount so utilized shall be required to be paid by the Purchaser at the closing of Title.

5. The occupancy of the Premises by the Purchaser shall be as a tenant at will.

6. Purchaser agrees to take the Premises in its "as is" condition and Seller shall have no obligation to make any repairs to the Premises or to remove any personal property from the Premises. The Seller's obligations, however, under the terms of the Purchase Agreement shall remain in full force and effect in accordance with the laws of the State of _____ and also the General Business Laws of the State of _____.

7. Neither this Agreement, nor the purchaser's taking possession of the Premises, shall be deemed a waiver of the Purchase Agreement but shall be considered as being conditional upon, and without prejudice to, the due performance of the Purchase Agreement in all respects.

8. All adjustments for taxes, water charges, and utilities shall be determined as of _____ and the Seller shall therefore be solely liable for same.

IN WITNESS HEREOF, the parties hereto have executed this Agreement as of the date and year first above written.

_____
(Community)

_____   by: _____
(Purchaser)   (Name, office)

_____
(Purchaser)

Where It's Filed

The early occupancy agreement is filed in the (prospective) purchaser's file folder.

PROTOCOL FOR A "COURTESY HOLD"

DESCRIPTION

The **Protocol for a "Courtesy Hold"** is a procedure for taking a reservation from a purchaser who is ready and willing but not able at the moment to leave a deposit or binder check.

PURPOSE

The purpose of the protocol for a courtesy hold is to accommodate a prospective purchaser who has completed a reservation to purchase a home with a short time period to provide a deposit or binder check. The prospect may have forgotten his/her checkbook and the community will not accept cash or credit cards.

USING THIS FORM

The protocol for a courtesy hold is used in the same way as the reservation form but it is valid only for a short time.

How To Use It

Before the prospective purchaser completes the reservation form, the sales counselor notes on it that it is valid only if a deposit is received by a certain date, usually not more than five days and rarely over the next weekend. The prospect may be given a stamped addressed envelope to mail the check back to the sales office if he/she doesn't intend to personally return with it.

The prospect should be advised that the community will accept other reservations on that home that will be retained in a secondary "back-up" position until the time expires for the receipt of the deposit check. If the prospect fails to meet the deadline, his/her reservation becomes null and void and the home will be offered to a back-up. If there are no back-ups, the home is placed back on the market and is available on a first come-first served basis.

This protocol should be used sparingly as a good faith gesture only in instances wherein the sales counselor seems certain of the prospect's sincerity and desire to proceed. If the sales staff records closings, contracts, and reservations on a color-coded site plan, a special color should be used to designate the courtesy hold.

CHAPTER 6

CLOSING PROCEDURES

OVERVIEW

The closing is the culmination of the formal, legal procedure in the acquisition of a new home. It should be a felicitous event and a relatively seamless one if each of the parties comes properly prepared. The builder, the construction staff, the sales staff, the purchaser, the lender, the attorneys, the title company, the surveyor, and the management company, if applicable, all have roles to play that must be coordinated in a timely fashion.

DESCRIPTION AND EXPLANATION

Planning and preparation for a closing entails many steps that require careful coordination and the assignment of specific responsibilities for each of the parties involved. The role of the builder's attorney and the sales staff will vary from community to community, so it becomes necessary to establish written guidelines and a checklist of items that need to be attended to. Appendix F offers a comprehensive closing preparations checklist. Most of the items are taken care of by the builder's attorney or in some cases by a specially appointed closing coordinator on the builder's staff. However, it is useful to have the sales staff prepare a closing cost financial schedule and a pre-closing letter for each purchaser. These forms are presented in this chapter.

HOW THE SYSTEM FITS WITH THE OTHER SYSTEMS

The closing procedures are the culmination of the sales process that began when the prospect entered the sales office and completed a prospect information card. Virtually all purchasers will require and welcome guidance and direction because investing in a new home is far from an everyday occurrence. Even experienced buyers may not recall the many details that require attention during what is generally a very hectic time. The closing cost financial schedule and the pre-closing letter are critical forms that will help to prepare the purchaser for the closing.

CLOSING COST FINANCIAL SCHEDULE

DESCRIPTION

The **Closing Cost Financial Schedule** is a detailed accounting statement and breakdown of the purchaser's total cost, payments, credits, and the balance due at closing. It itemizes the base price of the home, lot premium if applicable, purchases of options and upgrades, expenses, or credits incurred for change orders and payments made.

CLOSING COST FINANCIAL SCHEDULE

Date:

PURCHASER:_____ LOT #:_____STYLE:_____

SCHEDULED CLOSING DATE: To Be Determined
CLOSING SUMMARY

		AMOUNT	DATE PAID
1.	BASE PRICE OF HOME		
2.	DEPOSIT		
3.	LOT PREMIUM		
4.	UPGRADES SELECTED WITH CONTRACT		
5.	TOTAL CONTRACT		
6.	TOTAL DOWNPAYMENT (INCLUDING DEPOSIT)		

OPTIONS AND UPGRADES AFTER CONTRACT BY EXHIBIT A OR BY CHANGE ORDER

	AMOUNT	DATE PAID
EXHIBIT A:	$	
CHANGE ORDERS:		
1.	$	
2.	$	
3.	$	
TOTAL AMOUNT- EXHIBIT A OR CHANGE ORDERS	$	
TOTAL AMOUNT DUE	$	

CREDITS BY CHANGE ORDER

	AMOUNT
TOTAL AMOUNT CREDITS	$
TOTAL CREDITS DUE	$

CLOSING PRICE OF HOME	$
TOTAL PAID TO DATE	$
TOTAL CREDITS	$
TOTAL DUE AT CLOSING	$

Purchaser:_____Date:_____

Purchaser:_____Date:_____

PURPOSE

The purpose of the closing cost financial schedule is to provide the purchaser with an itemized bill so that the purchaser and his/her attorney will know in advance what funds are needed for the closing.

USING THIS FORM

Who Prepares It and Who Updates It

The sales staff or the closing coordinator prepares the closing cost financial schedule based on a review of the purchase agreement, the selections of options and upgrades, change orders, and other relevant correspondence of a financial nature in the purchaser's file folder.

Who Uses It

The sales staff, the purchaser, the attorneys, and the builder or the accounting department use the closing cost financial schedule.

How To Use It

Two copies of the closing cost financial schedule are sent to the purchaser for review about a month before the scheduled closing. If the purchaser believes the statement is in error, he/she will undoubtedly contact the sales office and the differences will be reconciled. If he/she believes it to be correct, the purchaser will return a signed copy to the sales office for their file folder and retain the other copy for his/her records. The sales staff will note the mailing out and the return of the closing cost financial schedule on the preliminary checklist for closings. A copy of the closing cost financial schedule will be forwarded to both the seller's and the purchaser's attorneys along with a copy of the Certificate of Occupancy and other relevant materials for the closing. The attorneys will use the closing cost financial schedule as a basis for the preparation of a more comprehensive closing statement (which might include adjustments for taxes, water and sewer fees, fuel costs, etc.) and for preparation of the RESPA statement.

How To Organize It

The closing cost financial schedule should be organized in a clear, concise format so that the purchaser can readily discern all of the financial elements of his/her transaction.

Who Gets Copies

The sales staff, the purchaser, the seller's attorney, the purchaser's attorney, the builder, and their accounting department should all receive copies of the closing cost financial schedule.

Method of Transmission

The closing cost financial schedule should be transmitted by mail or by fax.

Where It's Filed

The closing cost financial schedule is filed in the correspondence section of the purchaser's file folder.

PRE-CLOSING LETTER TO PURCHASER

DESCRIPTION

The **Pre-Closing Letter to Purchaser** is a checklist of items that need to be completed to assure and facilitate the closing.

PRE-CLOSING LETTER TO PURCHASER

Date

Mr. and Mrs._____

Address

Dear Mr. and Mrs._____

In order to close title and take possession of your new home as quickly and comfortably as possible, we remind you to attend to the following items:

1. Obtain your new address from the sales office.
2. Notify your post office and moving company. You should obtain change of address forms and a moving preparation checklist.
3. Check with your Lender to verify that ALL of the paperwork is completed and that you have satisfied ALL the conditions on your mortgage commitment.
   A If you have changed jobs since you initially applied, you must notify the Lender.
   B. Continue to exercise restraint in the use of credit card purchases until after the closing. A credit check will be run immediately prior to closing so that if you have built up large-scale debt, it may have to be paid off or you will lose your commitment.
4. Check with your attorney to verify the kinds and amount of checks (especially bank or certified checks) that you will need to bring to the closing along with any other required documentation. Note that unless arrangements have been made otherwise, all persons taking legal title must attend to and sign the mortgage documents.
5. Check with you insurance agent to ensure that you have the proper coverage required at closing.
6. Call the phone company (_____at_____) and cable

                     (Name)   (Number)

    company (_____at_____) to arrange for installation.

             (Name)   (Number)
7. Call the utility company (_____at_____) to arrange

                  (Name)   (Number)

    to have the power transferred to your name and account. Otherwise, the power will be turned off and you will incur a charge for resumption of service as well as endure an uncomfortable and inconvenient delay.

We hope that these reminders and suggestions are helpful during this exciting time and look forward to greeting you as our newest neighbor!

Sincerely,

PURPOSE

The purpose of the pre-closing letter to purchaser is to remind the purchaser of his/her final responsibilities and obligations that need to be met to be properly prepared for closing.

USING THIS FORM

Who Prepares It and Who Updates It
The sales staff prepares the pre-closing letter to the purchaser and updates it as warranted.

Who Uses It
The purchaser uses the pre-closing letter to the purchaser as a directive and guide in preparing for the closing.

How To Use It
The sales staff should send the pre-closing letter to the purchaser approximately four weeks prior to the projected closing date so that he/she has ample time to prepare for it. Its transmission should be noted on the preliminary checklist for closings.

How To Organize It
The pre-closing letter to purchaser should be organized as an itemized checklist for the purchaser to follow.

Who Gets Copies
The purchaser should receive a copy of the pre-closing letter to the purchaser.

Method of Transmission
The pre-closing letter to purchaser should be mailed to the purchaser.

Where It's Filed
The pre-closing letter to purchaser should be filed in the correspondence section of the purchaser's file folder.

CHAPTER 7

POST-CLOSING PURCHASER CONTACT

OVERVIEW

Post-closing purchaser contacts are too often neglected but they can be beneficial and instructive. We recommend that members of the sales and construction staffs jointly visit the new neighbor as soon as possible after the family has settled in. This is the ideal time to present a welcome gift, to explain customer service procedures and forms, to leave a list of emergency phone numbers, and to request that the purchasers complete a new homeowner's satisfaction survey and return it to the sales office (unsigned if they prefer).

DESCRIPTION AND EXPLANATION

This survey is designed to obtain the family's feedback on a variety of issues in addition to their satisfaction with their new home. These include the major reasons for their purchase; their evaluations of the sales staff, the construction staff, their mortgage officer, and the customer service department; and their recommendations for improving any aspect of the purchase process. Finally, the survey can be used to solicit the names and addresses of referrals that may be interested in moving to the community.

HOW THIS SYSTEM FITS WITH THE OTHER SYSTEMS

The new homeowner's satisfaction survey should be reviewed at least semi-annually to learn from the purchaser's perspective those areas that require attention and improvement and what adjustments and refinements should be made to enhance the community's sales, marketing, finance, and customer service programs. The respondents may also provide additional insight about their purchase decisions and the features of their homes, which may prove invaluable for the marketing program and for the delivery of future models or at least the incorporation of new features in older models. Finally, satisfied homeowners may also provide referral leads, a cost-free and felicitous source of potential new prospects that can be contacted to visit the community.

NEW HOMEOWNER'S SATISFACTION SURVEY

DESCRIPTION

The **New Homeowner's Satisfaction Survey** is a questionnaire about various aspects of the home-buying process.

NEW HOMEOWNER'S SATISFACTION SURVEY

Please take a few minutes to complete our survey. Your responses will help us to provide our homeowners with the highest quality of homes possible and to offer the highest form of customer service. When completed, please return the survey in the envelope provided to our sales office or fax it to us at _____

1. What have you enjoyed most about your new home? _____
    _____
    _____

2. What, if anything, would you like to change about your new home?_____
    _____
    _____

3. How would you evaluate our sales staff?_____
    _____
    _____

4. How would you evaluate our mortgage officer?_____
    _____
    _____

5. How would you evaluate our construction staff? _____
    _____
    _____

6. Did you use our service department? If yes, were your service requests taken care of promptly and to your satisfaction?_____
    _____

7. What were the three major factors in your decision to purchase a home in our community?
    (1) _____
    (2) _____
    (3) _____

8. Do you have any recommendations to make the purchase experience better? _____
    _____

9. Would you feel comfortable referring others to us to buy here? Yes_____ or No_____
    Why or why not?_____
    _____

10. Would you kindly provide the names of two friends who might enjoy living in one of our homes?

    Name(s)_____
    Address_____ City, State, Zip_____
    Phone number_____
    Name(s)_____
    Address_____ City, State, Zip_____
    Phone number_____

    The model of your home _____

    Your signature (optional)_____

    Thank you for your response. We appreciate and value your input.

PURPOSE

The purpose of the new homeowner's satisfaction survey is to solicit the opinions and recommendations of each purchaser so that improvements can be made to the design of future homes and so that the performance of various personnel and processes (sales, construction, finance, and customer service) can be evaluated. The new homeowner's satisfaction survey is also a means to solicit referrals.

USING THIS FORM

Who Prepares It and Who Updates It
The builder and his/her in-house staff should prepare this form and update it as needed.

Who Uses It
The builder and all affected company personnel.

How to Use It
The new homeowner's satisfaction survey should be distributed about two weeks after closing while events and impressions are still fresh in the purchaser's mind. A self-addressed stamped envelope should be included with a request that the survey be returned within a week. Signatures should be requested but are optional. Opinions vary about whether to combine the survey request with a visit to present a move-in gift or to explain customer service procedures or to treat it as an isolated event. Our preference is to combine it.

Either the in-house staff or a company that specializes in customer surveys should process the results. A summary of the findings should be circulated and discussed with company personnel to determine appropriate actions to correct deficiencies and achieve improvements.

How to Organize It
The new homeowner's satisfaction survey should be organized as a brief questionnaire that solicits customer feedback on specific areas of concern and also leaves space for general comments.

Who Gets Copies
The builder and all relevant staff members should receive copies.

Method of Transmission
The new homeowner's satisfaction survey should be personally delivered to each purchaser at his/her home.

Where It's Filed
Copies of the new homeowner's satisfaction survey should be individually filed in each purchaser's file folder and collectively in their own binder.

CHAPTER 8

COMMUNITY SHOPPING REPORT AND SALES PERFORMANCE EVALUATION

You never get a second chance to make a good first impression

OVERVIEW

Typical purchasers establish certain criteria, including location and price point limits, in their search for a new home. They will attempt to visit all of the communities that meet their requirements and pursue a policy of elimination before making what is usually the largest and most significant financial decision of their lives.

DESCRIPTION AND EXPLANATION

Because few if any new construction communities find themselves in the enviable position of having no competitors, it's essential that the sales and construction staffs visit competitive sites and obtain as much information as possible about each one. A community shopping report and sales performance evaluation form is invaluable for this purpose. It is a comprehensive critique and evaluation of a site, its sales process, and its sales counselor. The findings can be used in the preparation of one's own community comparison chart and to formulate responses to the inevitable comparisons that prospects will make with your community.

HOW THIS SYSTEM FITS WITH THE OTHER SYSTEMS

The community shopping report and sales performance evaluation is in effect a performance review that is carried out by every prospect who visits each community in the process of reaching a final decision in the search for a new home. Presentation is obviously critical. The community entry, the overall site appearance, the sales center, the model home(s), and the sales and construction staffs must be well prepared to promote a favorable impression. Knowledge of your own community and how it measures up against its competitors is indispensable to achieve success. Contrary to the popular saying "what you don't know can kill you," unless you are well informed about your competitors, your efforts sooner or later will be doomed to failure. This form will help to ensure the success of the community's sales program.

COMMUNITY SHOPPING REPORT AND SALES PERFORMANCE EVALUATION

DESCRIPTION

The **Community Shopping Report and Sales Performance Evaluation** is a critical analysis and comprehensive review of a competitor's community, the conduct of the sales process, and an evaluation of its sales counselors.

COMMUNITY SHOPPING REPORT AND SALES PERFORMANCE EVALUATION

Community:_____Type:_____Date:_____Time: _____

Phone #:_____Fax:_____Website:_____

E-mail:_____Day of Week:_____Weather:_____

Builder(s)/Developer/Sponsor: _____

Sales Consultant:_____Marketing Company:_____

Total # Residences Offered:_____No. Sold To Date:_____No. Unsold:_____

Total Acreage: No. of Home Choices:_____No. of Decorated Models:_____

Est. Delivery Time: Date of Pre-Opening:_____Date of Grand Opening: _____

Monthly Absorption Rate:_____Best Selling Model(s): _____

Amenities: _____

Positive Features: _____

Negative Features: _____

Unusual Features:_____

Special Incentives or Promotions:_____

_____

                                           Yes   No   Somewhat   Not Applicable

CONTACTING THE COMMUNITY

(a)   Was the phone number easily obtainable?_____

(b)   Was the sales center easy to reach by phone?_____

(c)   Did the respondent identify himself/herself?_____

(d)   Was your name and/or address requested?_____

(e)   Were your specific needs determined?_____

(f)   Were you encouraged to visit the community?_____

(g)   Was a specific appointment set up?_____

(h)   Did the respondent sound positive and professional?_____

(i)   Were directions offered?_____

(j)   Did they prove to be accurate?_____

ADVERTISING

(a)   Did you see an ad for this community?_____

(b)   Was it appealing?_____

(c)   Was the phone number accurate? _____

(d)   Were the directions correct?_____

                                           Yes   No   Somewhat   Not Applicable

LOCATING THE COMMUNITY

(a)   Did you notice directional signs en route?_____

(b)   Were they numerous enough, well-placed, and legible?_____

(c)   Was the community entry sign visible and attractive?_____

(d)   Did the community have "curb appeal"?_____

(e)   Was visitor parking convenient to the sales center?_____

THE SALES CENTER

(Located in: Trailer / Garage / Home (Circle One)

(a)   Was it easy to find?_____

(b)   Was it clean, neat, and professional in appearance?_____

(c)   Were provisions made for your comfort?_____

(d) Were refreshments offered?_____

(e) Were there attractive displays?_____

(f) Was a brochure offered?_____

(g) Were collateral materials offered?_____

THE GREETING

(By: Sales Manager / Sales Assistant / Host/Hostess (Circle One)

(a) Did you feel welcomed?_____

(b) Did the representative introduce himself/herself?_____

(c) Did the representative ask your name?_____

(d) Did the representative ask if this was your first visit?_____

(e) Were you asked what prompted your visit?_____

(f) Were you requested to complete a Welcome Card?_____

(g) If so, was it reviewed by the representative?_____

QUALIFICATION

(a) Was an effort made to determine why you're moving?_____

(b) Was an effort made to determine your specific needs?_____

(c) Was an effort made to determine your time frame?_____

(d) Was an effort made to determine your price range?_____

(e) Was an effort made to determine how long you've been looking for a new home?_____

(f) Was an effort made to determine what else you'd seen?_____

(g) If you named other communities, was the response to any of them:_____
   Positive_____
   Negative_____
   Not commented upon _____

Yes   No   Somewhat   Not Applicable

PRESENTATION

(a) Was the presentation concise and articulate?_____

(b) Was the presentation well organized?_____

(c) Was the information personalized to your concerns?_____

(d) Did the presentation sound mechanically rehearsed?_____

(e) Was an overview of the community presented?_____

(f) Were visual displays and collateral materials used?_____

(g) Was information on the builder/developer related?_____

(h) Was product knowledge evidenced?_____

(i) Were location advantages noted?_____

(j) Were features and their benefits explained?_____

(k) Were enthusiasm and pride evident?_____

DEMONSTRATION

(a) Were you accompanied on a tour of the model(s)?_____

(b) Were the models well maintained and appealing?_____

(c) Were the models identified by signs?_____

(d) Was a demonstration conducted?_____

(e) Were questions about suitability asked?_____

(f) Were your reactions asked for?_____

(g) Were your objections satisfactorily answered?_____

(h) Was the exterior as well as the interior included?_____

FINANCING

(a) Were you given financing information?

By: Sales manager _____ / mortgage officer_____ /

On: Pre-printed form_____ / personally prepared_____ /

In: Private_____ / Public_____ /

(b) Were your financial qualifications reviewed?_____

(c) Were alternative programs explained?_____

(d) Was a specific financial program selected?_____

(e) Were the tax benefits of home ownership explained?_____

(e) Was the mortgage process explained?_____

PROSPECT INVOLVEMENT
IN LOT & MODEL SELECTION

(a) Were available lots/units shown to you on a site map?_____

(b) Did the sales consultant assist you in selecting:

A specific model type or style of home or unit?_____

A specific home site or location?_____

(c) Did you walk the available lot(s)?_____

Alone_____

Escorted by the sales consultant_____

Escorted by construction department staffer_____

Yes   No   Somewhat   Not Applicable

CLOSING

(a) Were the steps in the purchase process explained?_____

(b) Did the sales consultant ask for a reservation deposit?_____

(c) Did the sales consultant create a sense of urgency?_____

(d) Did the sales consultant try to close more than once?_____

(e) Did the sales consultant show you a purchase agreement?_____

(f) Did the sales consultant offer a "courtesy hold"?_____

(f) Did the sales consultant attempt to schedule a return appointment at a specific time?_____

(h) Since your visit has there been site-initiated contact?_____

How soon afterwards_____ days

How many contacts?_____

What type(s) of contact?_____

THE SALES CONSULTANT

(a) Was the sales consultant's appearance professional?_____

(b) Was the sales consultant's manner professional?_____

(c) Was the sales consultant warm and friendly?_____

(d) Was the sales consultant enthusiastic and positive?_____

(e) Did the sales consultant try to establish rapport?_____

(f) Did the sales consultant successfully address your concerns and objections?_____

(g) Would you be comfortable working with the sales consultant?_____

(h) Did the sales consultant try to involve all of the family members?_____

(i) If you were the builder, would you be pleased to have the sales consultant
represent you and the community?_____

SUMMARY COMMENTS

_____

_____

_____

_____

PURPOSE

The purpose of the community shopping report and sales performance evaluation is to gain information about competitive communities to be able to prepare a community comparison chart and to serve your prospects with confidence since you will be as informed as they are and prepared to respond to their inevitable comparisons and questions.

USING THIS FORM

Who Prepares It and Who Updates It

The sales and construction staffs prepare this form and update it as warranted based on subsequent visits to the community, which should be scheduled on a regular basis approximately every six months.

Who Uses It

The sales and construction staffs use this form. One or both may want to modify their policies, procedures, or offerings to become or remain competitive based on the findings.

How to Use It

The community shopping report and sales performance evaluation should be completed during or soon after the visit to a competitive community while impressions are still vivid. The results should be reviewed and actions taken to enhance your competitive position. The findings can prove useful in preparing or improving your community comparison chart and in planning your responses to anticipated questions from prospects who have previously visited other sites (and will want to know why you don't offer certain items or do things like others do, or why your price per square foot is so much more, etc.).

How to Organize It

The community shopping report and sales performance evaluation should be organized into sections that take a visitor through a logical progression of steps from initial contact to the conclusion of his/her visit. It should include detailed questions about how the various steps in the "critical path" of sales were handled and the visitor's overall impressions of the community and the sales counselor. Space should also be included for general or summary comments.

Who Gets Copies

The sales and construction staffs should receive copies.

Method of Transmission

The community shopping report and sales performance evaluation should remain in the sales office.

Where It's Filed

The community shopping report and sales performance evaluations should be filed in their own binder.

Appendix A:
Sales Center Merchandising Checklist

Graphics Displays

A Welcome logo
A builder's panel
An area map
A site map and/or topo table
Elevations (renderings and floor plans)
A standard features list
Builder's professional awards and/or membership plaques
Pictures of the builder's previous communities
Lifestyle photos and/or an aerial view
Scale models
The current advertisement and/or public relations story
A bulletin board with community event photos
A TV with videotape about the builder, the community, and the general area

Equipment

A computer set-up
A typewriter
A two-line phone system with rollover number and directions at each phone location
Local phone directory and phone message pads and an answering machine
A fax machine with its own number
A copy machine and copy paper
A radio
An alarm system (for the models too)
Keys and "emergency key"
A postage meter or stamps
A microwave
A refrigerator
Photo boards for the construction team, testimonial letters, and realtors
A water cooler
Signs: available, sold, and out showing—please wait

Furnishings

Desk(s) and desk chair(s)
Desk accessories
Flags and pennants
Guest chairs and/or couches
Lamp(s)
File cabinets and hanging files
Table(s) and chairs
Plants

Sales documents and forms

Business cards
Community comparison chart
Registration/Welcome cards
Reservation forms
Cancellation forms
Realtor prospect registration forms
Co-broke protocols
Purchase agreements
Offering plans
Amendments
Community rules and regulations
W-9 forms
Change Order forms
Petty cash report and request forms
Brochures with inserts:
Price lists
Floor plans
Standard features list
Specification sheet(s)
Site maps
Builder information
Lot availability list with delivery dates
Warrantees
Architectural agreements
Special Use letters
Selections forms
Weekly Traffic and Sales Report forms
Weekly Tracking Report forms
Monthly Transactions (Sales Activity) Report forms
Thank you notes (for visiting)
Petty cash requisition forms

Collateral materials

Policy and procedures manual and pro forma manual
Community information book
Project photo album
A selections book with prices
Complete selections samples
Vendor catalogues
Chamber of Commerce materials
Finance information handouts
Credit check authorization forms
Train and bus schedules
School system information
Recreational facilities information
Binders for ads and PR stories and various reports
Comparable sales lists
Top Ten Prospects Lists
Job Shop Evaluation forms
Purchaser Profile forms

Customer Service Request forms
Post-Closing Survey forms
Schedule of Closings forms
Checklist For Closings forms
Closing Schedule Financial forms
Homeowner's directory
Community newsletter(s)
Specialty giveaway items

Office supplies

Restroom sundries
Kitchen and household sundries
A first aid kit
A cigarette urn
Stationery and envelopes
Stamps
Mailing labels and large envelopes
A rolodex with emergency and emergency warrantee phone numbers
A calendar
A day book (diary)
A daily planner/organizer
Clipboard(s)
A camera
A file box with dividers for registration cards
A broom and/or vacuum
A shovel and de-icer materials
A watering can
Extra light bulbs
File folders and labels for them
Hanging files and hanging file racks
Legal-size lined tablets
Manila folders (legal size and regular size) and manila folder labels
Pads
Paper clips
Two- and three-hole punch
Pens, pencils, markers, and white out
A pencil sharpener
A ruler
Rubber bands
A stapler, staples, and staple remover
Push pins and/or colored dots
Scissors
Scotch tape and dispenser
A tape measure
Wastepaper basket(s)
Plants
A welcome mat
A children's corner with coloring books, games, puzzles, toys, etc.

APPENDIX B: ACTIVITIES AND ROUTINES

An efficient sales office requires that certain activities be routinely carried out to insure its smooth operation. We recommend that the sales personnel use a daily planner to note appointments, calls, faxes and e-mail and priority items; a daybook to record significant policy or procedural changes (e.g. price changes, contract signings, closings, fall throughs, introduction of incentives, etc.); and a monthly calendar to note holidays, vacation schedules, and special events (e.g. Realtor's open house, opening of the clubhouse, introduction of a new model, homebuyer's seminar, etc.).

Certain tasks require daily attention while others may be attended to on a weekly or bi-weekly, monthly, annual or seasonal basis. We have listed those activities according to our preferences but each community will determine its own schedule.

DAILY ACTIVITIES

Drive to the site following the advertised directions checking for accuracy and noting any obstructions that may call for revision. Are the directional or "snipe" signs adequate in number and easily visible?

Inspect the physical status of the entry and site signage for the parking area(s), the sales center and model home(s), as well as the sold, available and lot number signs. Report any items requiring replacement or repair to the construction office.

Turn on the lights, turn off the alarm and check the answering machine and the fax machine in the sales center. Tidy up as necessary. Adjust heat or AC for maximum comfort. Put the coffee on.

Open the model(s), turn off the alarm, turn on the lights, note any items that require repair or touch up for forwarding to construction. Check the basement too.

Review the daily planner and monthly calendar.

Prepare and mail out purchaser correspondence, purchase agreements, closing statements, cancellation forms, thank you notes (for visiting the community) and other follow-up materials. Revise the top ten prospect list and take appropriate action.

Schedule selections appointments.

Add appropriate entries into the day book.

Check the mail.

Prepare new file folders for each new reservation.

Update the color-coded site map (with dots or push pins) to reflect new reservations, contracts, Closings, fall throughs and "courtesy holds."

Update the file folders as needed (e.g. adding the mortgage commitment).

Update the "comparables" after each closing.

Update the preliminary checklist for closings report.

Maintain a list of emergency phone numbers. (This is especially useful on the weekends when construction personnel are not usually available. E.g. pipe freezeup, air conditioning, heating or hot water failure, etc.).

Before leaving, turn the lights out, close all windows and doors, turn off and clean the coffee pot and set the alarm(s) in the sales center and model(s). Write entries into the daily planner for the next day. Be sure to turn down the heat and turn off the AC. Take a last look around before leaving and check to see that the door is locked.

WEEKLY AND BI-WEEKLY ACTIVITIES

Order office and rest room supplies as needed. At least a month's supply of all items is desirable. Allow more time to replenish printed materials.

Prepare the weekly traffic and sales report

Prepare the weekly tracking report

Update the top ten prospect list

Prepare agenda for the (weekly) construction/builder meeting. Transcribe the notes and distribute them to all attendees. Do the same for the sales meeting and be sure to focus on the top ten prospect list.

Cut out copies of the advertisements and retain them in a separate binder.

Prepare a sufficient number of complete brochures for the coming weekend and the following week.

Report any deficiencies in the janitorial service.

Every Friday, prepare materials and the sales office environment and rest rooms for the weekend. Update the lot availability list and confirm with construction which homes being built can be shown, along with the models, over the weekend. Are special keys required for access?

MONTHLY ACTIVITIES

Prepare the transactions (sales activity) report

Review and renew Realtor contacts

Send out a list of monthly sales to whoever requires it (e.g. Management Company, Municipal agencies, etc.)

ANNUAL ACTIVITIES

Prepare the annual report.

Review last year's annual report. Were goals met? Projections achieved?

Review the purchaser family profile forms, selections and new home owner satisfaction surveys to determine whether product refinements, changes in procedures or revisions to the Advertisements and marketing are warranted.

Check on continuing education real estate licensing requirements.

SEASONAL ACTIVITIES

In winter, be sure to have a shovel and an adequate supply of ice melt, salt or sand.

You may have to remove snow and ice from the sidewalks and entry areas to the parking area, sales center and model home(s).

During the season, be sure that all grassed and landscaped areas are adequately watered and appropriately trimmed.

Change the answering machine message as necessary. (eg. closed for the Holidays). Place a sign outside the sales office too. You might install outdoor holders with brochures and prospect information cards available at times the sales office is closed.

APPENDIX C: SETTING UP A FILE FOLDER SYSTEM

A individual file folder for each lot or unit should be opened as soon as a reservation form is completed and regularly updated thereafter to document and monitor each step in the process and the progress of the sale. The file folders should be arranged numerically and stored collectively for easy reference.

We recommend the 11¾ × 10 inch expandable pressboard classification folders. They include two interior partitions and six fasteners to hold two hole-punched materials in place. The inside front cover holds the purchaser information card and the family profile form. The first partition retains the reservation form, a W-9 form, the memorandum of agreement and copies of all checks. Its reverse side is for communications, correspondence and notes. The second partition binds the selection sheets, a copy of the floorplan and any change orders. Its reverse side is for a copy of the loan commitment. The back inside cover retains a copy of the purchase agreement. The attorney's closing statement is the final document to be included. Thus each file folder embodies the complete history of its transaction.

File folders for all closed transactions and fall-throughs should be retained and stored separately. Labeling the raised tabs of the file folders with the purchaser's name, lot or unit number and model type or style of the residence provides quick and easy identification and retrieval.

In addition to individual file folders for each lot or unit, the sales center should also maintain a master file of all current administrative and legal forms, an elevation, floor plan and scale drawings file, an interoffice correspondence file, an advertising and public relations file and an archive file of all old originals. Individual binders for the regular reports and checklists itemized in the previous pages should also be maintained and kept readily accessible.

APPENDIX D: SOME OF THE MOST FREQUENTLY ASKED QUESTIONS IN THE SALES OFFICE

As part of the Critical Path role-play, the sales staff should have prepared responses to the inevitable visitors' frequently asked questions (FAQS).

1. Is the model unit/home for sale?
2. Who furnished the models or supplied the mirrors or levelers or built-ins, etc. (Can I get these items)?
3. Are there any upgrades or options available?
4. Will I receive an allowance if I do not accept what is offered in your standard package? (I have my own brand new refrigerator.)
5. Why is the credit for a refrigerator so little? I checked the cost at_____. It's much more there and elsewhere too.
6. Can a humidifier or dehumidifier be attached to the heating system?
7. Can I install a skylight? Track lighting?
8. Will a king size bed fit in your master bedroom?
9. Will you install flooring in the entire attic storage area and/or add extra insulation?
10. Is the foundation poured or block?
11. What type of storm door must I have? Do I have a choice?
12. Will you break through the wall of an adjacent unit if I buy two adjoining units?
13. Can I get a reserved or handicapped parking space close to my unit?
14. Can I fence in my home?
15. Can I add my own shrubbery? Garden?
16. What are the rules and regulations in this community?
17. How long have you been selling here? Has there been a price increase?
18. Can I use my home/unit as a professional office?
19. Do you allow pets? If my pet dies, can I replace it?
20. What kind of people live here? (Direct inquiries concerning race, religion, etc.)
21. Is the price negotiable?
22. What kind of financing do you offer?
23. Is there a discount for an all-cash purchase? (If not, why not, especially since you are willing to subsidize some closing costs?)
24. What happens to my downpayment? Does it earn interest? If so, who is it credited to?
25. Will you accept an assignment of a CD/T-bill/money market certificate for the downpayment?
26. Will you accept a "staggered" downpayment schedule?
27. What is included in the monthly HOA fee or common charge fee?
28. Will you itemize our closing costs?
29. What's the total monthly outlay and net monthly cost?
30. What can I expect to rent my unit for? Will you find me a tenant?
31. Is my purchase agreement assignable?
32. What is the commute time and cost to _____?
33. Is there parking at the station? Cost? Waiting list?
34. What is the board of managers/directors?
35. What kind of insurance do I need?

36. Do you have information on schools, hospitals, religious institutions, shopping, area amenities (community pool, golf course, etc.)?
37. Does the local school system offer special education courses or provide classes for autistic children or for advanced placement?
38. Can I get a larger deck?
39. Can you remove the posts in the basement and replace them with steel beams?
40. Why is the washer/dryer upstairs? If it overflows won't there be damage?
41. Can I move the washer/dryer? (to the basement?)
42. Can I put a half bath or full bath in the basement?
43. Are screens included?
44. What type of warranties do you offer?
45. Tell me about your builder? How long has he/she been in business? Where else has he/she built? Any references you can give me?
46. Can you give me the names and numbers of some homeowners?
47. What happens if my mortgage rate-lock expires before the home is ready for delivery? If there's a fee for an extension, who pays it?
48. Can I move in or move my furniture in before I close?
49. Can I use your bathroom? Phone? Can I get a drink?
50. Oh, I'm working with a realtor but I found this place on my own. Will you honor my realtor?
51. What do we do? My son locked himself in the bathroom!
52. It's Saturday at 4:50 and you close at 5 P.M. The phone rings. "We're on our way. We got lost. We've been driving for an almost two hours; we'll be there in 20 minutes. Will you wait for us"?

Appendix E: Alternative Lifestyles, Condominiums, Co-operatives, and Homeowners Associations

Increasingly, condominiums, co-operatives, and homeowners association units are becoming the preferred lifestyle for people of all ages and economic groups. No longer are these residences looked upon as housing for the "newlywed or the nearly dead." Let's consider each one.

A condominium apartment or "unit" is real property. The purchaser receives a recordable deed just as if he/she were buying a single-family home or a parcel of land. Each unit is separately taxed and each buyer is responsible for his/her own real estate taxes. In addition, there is a separate monthly fee, called a common charge, that is usually based on the size of the unit and is used to run and maintain the condominium complex. The monthly common charge is similar to the monthly maintenance charge in a co-operative. Common charges are not tax deductible and do not include real estate taxes. The condominium is governed by an elected board of managers who administer its affairs and enforce its by-laws and rules and regulations. The board may have a "right of first refusal" on sub-leasing or sales.

The condominium may be an all-cash purchase or the buyer can apply for a mortgage and may be able to finance 90% or even 95% of the purchase price if qualified to do so. Conventional financing with standard bank closing costs is readily available and there is no board interview required as is the case with a co-operative. Given that they are easy to purchase and that the monthly combined common charges and real estate taxes are generally less than a co-operative's monthly maintenance fees, condominiums are generally more expensive than co-ops. Additionally, condos are less expensive to carry than a co-op because there is no underlying mortgage on a condominium building or development. The condominium owner owns the interior of his unit, not the exterior or the land under (except in the state of Michigan) or around his dwelling. The exterior and the land and any facilities are "common elements" or "common areas," and each condo owner has an "undivided percentage interest" in such areas. Finally, the term "condominium" does not refer to a specific form of construction but instead to a unique method of ownership.

Co-operatives are a type of ownership more common in New York City than elsewhere in the United States, where they constitute approximately 85% of the apartments available for purchase. Co-operatives are owned by an apartment corporation. When you purchase shares within a co-operative building, you become what is known as a "shareholder," which entitles you (the shareholder) to a long-term "proprietary lease." The individual shareholder or stockholder does not actually "own" his/her apartment but a percentage of the total shares within the co-operative. The larger the co-operative apartment, the higher the percentage of ownership in the corporation and, accordingly, the higher the monthly maintenance fee.

Co-operative shareholders pay a monthly maintenance fee to cover the expenses of their building. This fee covers heat, hot water, insurance, mortgage indebtedness, real estate taxes, and staff salaries. Most co-operative corporations have underlying mortgages on their buildings. A portion of the monthly maintenance contribution (for taxes and the building's underlying mortgage interest) as well as the interest on the co-op loan is tax deductible. If a shareholder defaults on his/her monthly payment, the remaining shareholders are liable.

All prospective co-operative apartment purchasers are interviewed by members of the board of directors who have the right to approve or disapprove the applicant. There is a minimum downpayment set by the board, which is usually a minimum of 20 to 25 percent in cash. Because a personal loan and not a mortgage is involved (unless it's an all-cash transaction), the closing costs are far less than in a condominium. Each co-operative has its own rules that should be carefully reviewed prior to an application for purchase.

Sub-leasing provisions can be onerous. Unanticipated building repairs or improvements may cause the board to levy "special assessments" on the shareholders. Finally, when you decide to sell, the prospective purchaser must be approved by the board of directors. There is more support for the quality of life and security of the building than in most condominium communities.

The homeowners association is somewhat of a hybrid form of ownership, ideally combining the best features of single-family and condominium ownership. It is a legal form of fee-simple ownership in which the individual owns his/her own detached home or attached townhome (a term without statutory definition) in a community with one or both styles of residence and the land on which it is set while an association (consisting of all owners) owns the common areas, which usually include recreational facilities and amenities. As with the condo and co-op, there are rules and regulations set forth in an Offering Plan or Prospectus, which is a disclosure document required by state law, to sell the residences. There are rules and regulations and a board of directors to administer and manage the affairs of the HOA. The following chart illustrates some of the differences and similarities between the three types of alternative lifestyles:

ALTERNATIVE LIFESTYLE CHART

	Condominium	Co-operative	Homeowners Association
Offering by:	Prospectus/Plan	Prospectus/Plan	Prospectus/Plan
Includes:	Declaration By-Laws Rules & Regulations	By-Laws Rules & Regulations	By-Laws Rules & Regulations
Ownership:	Fee Simple (Interior of unit and % interest common elements)	Leasehold (% interest common elements)	Fee Simple (interior and exterior of and land, associated common elements)
Documents:	Deed	Proprietary Lease Membership Certificate (Shares of Stock)	Deed
Purchase:	Cash/Mortgage	Cash/Personal Loan	Cash/Mortgage
Financing:	Conventional/Market Rate	Generally, higher downpayment and interest rate	Conventional/Market Rate
Closing Costs:	Standard Bank	No Bank Closing Costs	Standard Bank
Common Charges	For Maintenance	For Maintenance	For Maintenance
HOA Fees		For pro rata share of taxes and underlying mortgage(s)	
	Not Tax Deductible	Taxes and interest on mortgage are deductible	Not Tax Deductible
Government:	Board of Directors / Manager	Board of Directors / Manager	Board of Directors / Manager
Lease/Resale Provisions:	Often, right of first refusal; specific legal forms: waiver of the right of first refusal, certificate of insurance, common charge lien letter, power of attorney	Often, interview; specific legal forms: waiver of the right of first refusal, certificate of insurance, common charge lien letter, power of attorney May have to sell back to the board with little or no profit or pay a specific fee to the board	Generally, none; rentals maybe for a minimum period and to a maximum number of people
If Default in Monthly C.C.	No effect on other owners	Other leaseholders must "chip in"	No effect on other owners
Positive features for all:	tax benefits, equity build up, appreciation potential, maintenance done by professionals, usually recreational amenities, mini-government, sense of community		

Negative features may include (especially for condominiums and cooperatives): Loss of freedom and privacy, rules and regulations (from pets to parking), limits on use of units/apartments, competency of board of managers/directors

Note: In a condominium, the declaration and by-laws and rules and regulations "survive" the closing. They are require for re-sales.

A useful comparison of the benefits of owning versus renting is providing in the chart that follows.

A QUICK COMPARISON BETWEEN OWNING A CONDOMINIUM VERSUS RENTING, VERSUS OWNING A CO-OPERATIVE, VERSUS OWNING A PRIVATE HOME

	Condominium Ownership	Rental Apartments	Co-operative Apartments	Private Home
1. Tax savings	Yes	No	Yes	Yes
2. Equity build up	Yes	No	Yes	Yes
3. Appreciation and profit potential	Yes	No	Yes	Yes
4. Stabilized monthly costs due to fixed principal and interest payments	Yes	No	Yes	Yes
5. Individual conventional mortgage financing readily available	Yes	No	No	Yes
6. Individual ownership of your residence	Yes	No	No	Yes
7. Freedom of responsibility from your neighbor's default	Yes	Yes	No	Yes
8. Maintenance done for your by professionals	Yes	Yes	Yes	No

APPENDIX F: CLOSING PREPARATION CHECKLIST

◆ Determine target date with construction department for the CO

◆ Set up a closing date by phone with the purchaser

◆ Confirm date with purchaser at least 10 days prior to closing by certified mail

◆ Confirm date and time with lending institution and purchaser's attorney and our attorney

◆ Prepare and mail out purchaser's closing cost financial statement

◆ Send signed purchaser's closing cost financial statement to our attorney

◆ Send brokerage fee bill to our attorney (if applicable)

◆ Obtain copy of inspection report from construction department

◆ Prepare "survival letter" i.e. items to be completed after closing

◆ Check to see that release of mortgage was ordered

◆ Check with title company to see if they received release of mortgage

◆ Check to see that the survey has been received and certification is accurate.

◆ Request mortgage release check plus any other necessary checks

◆ Prepare the deed

◆ Prepare the power of attorney

◆ Prepare mortgage tax affidavit

◆ Complete certificate of insurance

◆ Send copies of certificate of insurance to insurance carrier

◆ Send copy of building's flood insurance (if applicable)

◆ Supply copy of the resolution for the President to sign the deed

◆ Have the President sign the deed

◆ Seal the deed with the corporate seal

◆ Notarize the President's signature

◆ Have the President sign the mortgage tax affidavit

◆ Notarize the President's signature

◆ Have the mortgage release check and transfer tax check, if applicable, to give to the title company at the closing

◆ Have a copy of the release available if not closing with the usual title company

◆ Have a check for the title company (if not the usual title company) for conveyance tax, mortgage tax, if any, and fee for release of mortgage

◆ Have copy of CO for lender if not mailed out in advance

◆ Have a copy of home owner's maintenance manual

◆ Bring keys, key card, mailbox key, etc. if the procedure calls for this at closing

◆ Bring checks for any municipal taxes due.

Note: Always bring the complete property file folder, an offering plan and all amendments (if applicable) as well as a calculator and legal pad. Know where your builder can be reached in case of any unforeseen problems that may arise.

FORM LIST

Form	Purpose	Chapter
Prospect Info Card	This form is used to obtain basic demographic information about the prospect and their housing requirements.	1
Community Comparison Chart	This form is a sales aid that differentiates the community from its competition by highlighting its appealing, valuable and unique features. (Not available on disk.)	1
Inventory Availability List	This form is designed to control the inventory and provide for an orderly pattern of sales in the construction of the community.	1
Co-Brokerage Protocol	This form is designed to detail how a builder can increase their community's sales by implementing an effective and mutually beneficial Co-op Program with the real estate community.	1
Realtor Prospect Registration and Confirmation Form	This form activates and implements the Co-Brokerage Protocol for a specific sales transaction.	1
Reservation Form	This form details the agreement between the purchaser and the builder (or Sponsor of a co-operative, condominium or Home Owner Association unit) so that they can formalize their intentions in a Purchase Agreement.	1
Cancellation Form	This form is used to record the reason(s) the purchaser is unwilling or unable to proceed with the transaction and to determine what, if anything, the sales counselor or builder can do to salvage the sale.	1
Memorandum of Agreement	This form is a summary of the basic facts of a sales transaction.	1
Purchase Agreement	This form is a legal document containing the terms and conditions for the conveyance of title to real property from a seller to a purchaser.	1
Architectural Agreement	This form allows the builder to provide the purchasers with the option of making custom changes.	1
Purchaser Family Profile Form	This form is designed to collect data about each family member including place(s) of employment, finances and financing relevant to the purchase and information about the search for a new residence.	1
Weekly Traffic and Sales Report	This form is used to monitor the community's traffic and sales activity on a daily basis with weekly summations.	2
Top Ten Prospect List and Follow-Up Form	This form serves as a reminder to schedule immediate and continuous contact with ready, willing, able and interested prospects until a purchase decision is made.	2
Weekly Tracking Report	This form monitors the progress of every sale from the reservation to the closing and provides for significant comments and key party contact information.	2
Monthly Transactions Report	This form provides a concise summary of the site's sales activity on a month to month, year-to-date and overall basis from the inception of the sales program to the most recent month's activity.	2
Schedule of Future Closings Report	This form enable the sales and construction staffs to monitor the construction progress of each home so that its completion is on schedule to meet the delivery date specified in its Purchase Agreement.	2
Preliminary Checklist for Closings	This form is used to remind the sales staff to take the necessary steps and secure the appropriate documentation to insure a timely closing.	2
Comparable Sales Report	This form is used to facilitate the work of appraisers and thereby expedite the mortgage approval and commitment process.	2
Annual Report	This form provides information for analysis of the achievements and shortcomings of the past twelve months, to set goals for the future and to determine what changes should be implemented to enhance performance and increase bottom line profitability. (Not available on disk.)	2
Protocol for Selections, Options, and Upgrades	This form provides a complete record of all the standard features, options, upgrades, modifications (additions, deletions and substitutions) selected and the charges or credits that apply.	3
Information/Price Request Form	This form is used to obtain a written response from the construction department on concerns or inquiries posed by purchasers.	3
Change Order Form	This form is used to record a purchaser's custom changes to their residence (modifications, additions, deletions, or substitutions), the cost involved, and the terms and conditions under which approved changes will be made.	3

GLOSSARY

addendum. Special addition to a form or purchase agreement that affects the total form or purchase agreement. Also known as a rider.

agent. Someone who is authorized to represent another person.

agreement of sale. Written agreement between seller and purchaser in which the purchaser agrees to buy certain real estate and the seller agrees to sell on terms and conditions set forth therein. The purchaser leaves a deposit or partial downpayment with the balance due at the signing of a contract or purchase agreement. Also known as a sales agreement.

appraisal. Formal opinion or estimate of value by one who is qualified (an appraiser) to evaluate factors of value. Appraisals are based on style and appearance, construction quality, usefulness, and recent sales of comparable properties.

binder. Agreement to cover the downpayment for the purchase of real estate as evidence of good faith on the part of the purchaser.

broker. One who acts as an agent or negotiator for his principal when dealing with third parties.

building envelope. That portion of a building lot within which all of the elements of the building must be enclosed or confined to.

certificate of occupancy. Document issued by a zoning board, building department, or other municipal agency to indicate that a structure complies with building code requirements and may be legally occupied.

change order. Order to change the work to be performed under a construction contract. Usually the purchaser will request a change that the sales staff will submit to the construction staff for approval and pricing.

client. Principal to a real estate transaction who employs an agent.

closing. Final step where ownership of a residence is transferred from the seller to the purchaser. Also known as the settlement.

closing costs. Fees and other charges paid by both the buyer and the seller at the closing.

common charges. Monthly maintenance fee that condominium owners pay. Analogous to the monthly maintenance fee that co-operative owners pay although a portion of that fee is tax deductible since it also includes payment for taxes and a pro-rata share of the building's underlying mortgage.

condominium. Form of real estate ownership in which the owner receives title to a particular unit and has a proportionate, undivided interest in certain common areas and facilities. The unit itself is generally a separately owned space whose interior surfaces (ceilings, floor, and walls) serve as its boundaries. Dwelling units are mortgaged separately, and each unit owner contributes a share of the expense for the maintenance and operation of common facilities, which are administered by an elected board of managers. See Appendix E.

contract. Agreement between competent parties to do or not to do certain things that is legally enforceable and whereby each party acquires a right. A contract of sale in real estate is also known as a purchase agreement or sales contract.

cooperative. Form of multiple ownership in a housing corporation owned by residents and operated for the benefit of resident members of the corporation by their elected board of directors. The resident occupies but does not own his unit. Instead, he/she owns a share of stock or membership certificate in the total enterprise and has a proprietary lease. See Appendix E.

critical path. Series of activities and events designed to take a prospect in a logical progression of steps from the initial greeting to the signing of a purchase agreement and follow-up after the closing.

deed. Instrument in writing duly executed and delivered that conveys title to real property.

escrow. Written agreement between two or more parties providing that certain instruments or property be placed with a third party to be delivered to a designated person on the fulfillment or performance of some act or condition.

Fannie Mae. Popular name of the Federal National Mortgage Association; originally created to provide a degree of liquidity in the mortgage market by establishing a secondary market for existing mortgages. The agency does not loan money directly but buys mortgages with bond-generated funds, which are originated by other lending institutions. It is the nation's largest purchaser of home mortgages.

fee simple. Absolute ownership of land with no restrictions on the transfer of ownership.

FHA Loan. Loan insured by the Federal Housing Administration (of the Department of Housing and Urban Development).

Freddie Mac. Popular name for the Federal Home Loan Mortgage Corporation. FHLMC buys mortgages in the secondary market from commercial banks having insured deposits or from federally insured savings and loan associations belonging to the Federal Home Loan Bank System. It includes FHA and VA loans as well.

gift letter. Letter from a donor to an applicant for a mortgage stating the donor's name and address, the amount of the gift, and a statement that the gift does not have to be repaid or returned. The gift may be required for the applicant to qualify for a mortgage from the lender.

grace period. Additional time allowed to perform an act or make a payment before a default occurs.

homeowner orientation. Final walk-through inspection of a property's condition conducted by the purchaser (accompanied by the builder or a member of the builder's construction staff) to ensure that all of the conditions noted in the purchase agreement, and the selections and change orders, have been met.

homeowners association residence. Form of ownership in which a resident owns his/her home and pays a monthly fee for the maintenance of common areas or facilities. See Appendix E.

mortgage. Instrument in writing, duly executed and delivered, that creates a lien on real estate as security for the payment of a specified debt, which is usually in the form of a bond.

mortgage commitment. Formal indication by a lending institution that it will grant a mortgage loan on property in a certain specified amount and on certain specific terms.

Multiple Listing Service (MLS). System that provides its members with detailed information about property for sale and arranges for co-operating members to share brokerage fees where warranted.

offering plan. Legal document, usually in the form of a bound book, that serves as a disclosure document drawn up in accordance with state law and required to sell newly constructed condominiums, co-operatives, or homeowners association residences as well as condominium or cooperative conversions of rental units. Also known as a prospectus.

prospect. Potential purchaser who has indicated at least minimal interest and who is known to the sales staff by name, address, and telephone number.

rate-lock. Interest rate that is locked in or guaranteed for a specific time.

realtor. Coined word that may only be used by an active member of a local real estate board affiliated with the National Association of Realtor Boards.

reservation. Agreement made preliminary to the contract of sale or purchase agreement to sell property, used as a temporary arrangement before entering into the formal contract. It requires a deposit given in good faith to provide evidence of ability and willingness to complete the purchase of real property.

Real Estate Settlement Practices Act. RESPA. A federal law enacted in 1975 to force disclosure of all aspects of financing to potential borrowers.

setback requirements. Distance from the curb or other established line, within which no buildings may be erected. Municipalities generally also specify sideyard and backyard requirements as well.

stipulations. Terms within a written contract.

survey. Process by which a parcel of land is measured and its boundaries ascertained. The actual document with such measurements, which, at time of closing, also includes the certifications (the names of the purchasers, their lender, and title company and policy information).

title. Evidence of ownership and lawful possession of property, normally in the form of a deed.

title search or examination. Check of the title records and legal proceedings, generally at the local courthouse, to make sure that you are buying a dwelling from the legal owner and that there are no liens, overdue special assessments, other claims, outstanding legal covenants, or other defects in the title filed in the records.

VA loan. Loan made by a private lender that is partially guaranteed by the Veterans Administration.

SELECTED BIBLIOGRAPHY AND RECOMMENDED REFERENCES

Alfriend, Bonnie and Tiller, Richard. *New Home Sales Management.* Pebble Beach, CA. Alfriend and Associates, 1996.

Asdal, William and Jordan, Wendy A. *The Paper Trail: Systems and Forms for a Well-Run Remodeling Company.* Washington, DC. BuilderBooks, 2002.

Clark, Charles R. and Parker, David. F. *Marketing New Homes.* Washington, DC. National Association of Home Builders, 1989.

Clark, Charles R. and Parker, David F. *Selling New Homes.* Washington, DC. National Association of Home Builders, 1989.

Covey, Stephen R. *The Seven Habits of Highly Effective People.* New York. Simon and Schuster, 1989.

Davenport-Ennis, Nancy. *New Home Selling Strategies.* Chicago. Dearborn Financial Publishing, 1992.

Drucker, Peter F. *Management: Tasks, Responsibilities and Practices.* New York. Harper and Row, 1974.

Mitchell, Jan. *Sales and Marketing Checklists for Profit-Driven Home Builders.* Washington, DC. BuilderBooks, 1997.

National Association of Home Builders. *Home Builder Contracts & Management Forms on Disk.* Washington, DC. 2001.

National Association of Home Builders. *Increased Profits Through Effective Builder Broker Co-Operation.* Washington, DC. 1999.

National Association of Realtors Website (www.realtor.org) Library Leads Field Guide. Selling **New Homes** for compilation of useful materials.

Peters, Thomas J. and Waterman, Robert H. Jr. *In Search of Excellence.* New York. Warner Books, 1982.

Radice, Dennis. *The Home Builders Sales and Management Tool Kit: Working with Brokers, Agents and Onsite Associates.* Washington, DC. BuilderBooks, 2000.

Smith, Carol. *Dear Homeowner, A Book of Customer Service Letters.* Washington, DC. BuilderBooks, 2000.

Smith, Carol. *Meeting with Clients: A Self-Study Manual for a Builder's Frontline Personnel.* Washington, DC. BuilderBooks, 2002.

Stone, David. *New Home Sales.* Chicago. Longman Group, 1982.

Stone, David. *Policies and Procedures Manual for New Home Sales.* Los Gatos, CA. The Marketing Forum, 1984.

Stone, David and Ritchey, Tom. *Role of the Sales Manager.* (video film) Washington, DC. National Association of Home Builders, 1986.

Stone, David. *Community Control Manual* (manual). Los Gatos, CA. The Marketing Forum, 1987.

Stone, David. *New Home Marketing.* Chicago. Longman Group, 1989.

Walsh, Dennis and Associates. *Certified New Home Specialist.* (video) Newport Beach, A 2001. Also companion videos *Presentation Builder and Toolbox.*

Winston, Stephanie. *The Organized Executive.* New York. W. W. Norton, 1983.